P9-CJT-066

New Security Challenges Series

General Editor: **Stuart Croft**, Professor in the Department of Political Science and International Studies at the University of Birmingham,UK

The last decade demonstrated that threats to security vary greatly in their causes and manifestations, and that they invite interest and demand responses from the social sciences, civil society and a very broad policy community. In the past, the avoidance of war was the primary objective, but with the end of the Cold War the retention of military defence as the centrepiece of international security agenda became untenable. There has been, therefore, a significant shift in emphasis away from traditional approaches to security to a new agenda that talks of the softer side of security, in terms of human security, economic security and environmental security. The topical *New Security Challenges series* reflects this pressing political and research agenda.

Titles include:

Brian Rappert
BIOTECHNOLOGY, SECURITY AND THE SEARCH FOR LIMITS
An Inquiry into Research and Methods

Brian Rappert *(editor)*
TECHNOLOGY AND SECURITY
Governing Threats in the New Millennium

New Security Challenges Series
Series Standing Order ISBN 0–230–00216–1 (hardback) and ISBN 0–230–00217–X (paperback)

You can receive future titles in this series as they are published by placing a standing order. Please contact your bookseller or, in case of difficulty, write to us at the address below with your name and address, the title of the series and one of the ISBNs quoted above.

Customer Services Department, Macmillan Distribution Ltd, Houndmills, Basingstoke, Hampshire RG21 6XS, England

---

Also by Brian Rappert

CONTESTED FUTURES (*co-editor with Nik Brown and Andrew Webster*)

CONTROLLING THE WEAPONS OF WAR: Politics, Persuasion and the Prohibition of Inhumanity

NON-LETHAL WEAPONS AS LEGITIMIZING FORCES?: Technology, Politics and the Management of Conflict

Biotechnology, Security and the Search for Limits

An Inquiry into Research and Methods

Brian Rappert
University of Exeter

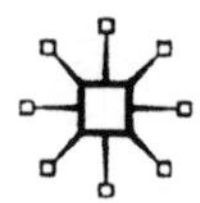

First published in 2007 by
PALGRAVE MACMILLAN
Houndmills, Basingstoke, Hampshire RG21 6XS and
175 Fifth Avenue, New York, N.Y. 10010
Companies and representatives throughout the world.

PALGRAVE MACMILLAN is the global academic imprint of the Palgrave Macmillan division of St. Martin's Press, LLC and of Palgrave Macmillan Ltd. Macmillan® is a registered trademark in the United States, United Kingdom and other countries. Palgrave is a registered trademark in the European Union and other countries.

ISBN-13: 978–0–230–00248–7 (hardcover)
ISBN-10: 0–230–00248–X (hardcover)

This book is printed on paper suitable for recycling and made from fully managed and sustained forest sources. Logging, pulping and manufacturing processes are expected to conform to the environmental regulations of the country of origin.

A catalogue record for this book is available from the British Library.

Library of Congress Cataloging-in-Publication Data

Rappert, Brian.
Biotechnology, security and the search for limits : an inquiry into research and methods / Brian Rappert.
p. cm. —(New security challenges)
Includes bibliographical references and index.
ISBN-10: 0–230–00248–X (cloth)
ISBN-13: 978–0–230–00248–7 (cloth)
1. Biotechnology – Social aspects. 2. Biotechnology – Risk assessment. 3. Biological arms control. I. Title.

TP248.23.R37 2007
660.6—dc22 2006052714

10 9 8 7 6 5 4 3 2 1
16 15 14 13 12 11 10 09 08 07

Printed and bound in Great Britain by
Antony Rowe Ltd, Chippenham and Eastbourne

Contents

Boxes and Table

Acknowledgements

The research undertaken as part of this book was funded by the UK Economic and Social Research Council (ESRC) New Security Challenges Programme (RES-223-25-0053) and approved by a research ethics committee at the University of Exeter. In addition, a grant from the Alfred P. Sloan Foundation enabled the author the space to undertake a more systematic analysis of educational themes. Presentations relating to the book's arguments were given at meetings held by the US National Academies, the Organization for the Prohibition of Chemical Weapons and the International Union of Pure and Applied Chemistry, the United Nations Department on Disarmament Affairs, the British Royal Society, the European Association for the Study of Science and Technology as well as at Exeter, Edinburgh, Arizona State, and Open Universities. My thanks to all those that participated in these events and provided valuable comments.

The process of inquiry elaborated could not have been undertaken without the assistance of many individuals. My particular thanks to Stuart Croft, David Dickson, Chandré Gould, Nick Green, Alastair Hay, Alan Malcolm, Kathryn Nixdorff, and Ben Rusek for their counsel. Harriet Tillson made helpful editorial comments throughout the book. Finally, of course, it was a great pleasure and benefit to undertake the life science seminars with Malcolm Dando.

Abbreviations

ASM	American Society for Microbiology
BTWC	Biological and Toxin Weapons Convention
CDC	Centers for Disease Control
CWC	Chemical Weapons Convention
DoD	Department of Defense
DHHS	Department of Health and Human Services
ESRC	Economic and Social Research Council
IAP	InterAcademy Panel
IUBS	International Union of Biological Sciences
ICSU	International Council for Science
IUBMB	International Union of Biochemistry and Molecular Biology
IL	Interleukin
NAS	National Academy of Sciences
NIAID	National Institute of Allergy and Infectious Diseases
NIH	National Institutes of Health
NSABB	National Science Advisory Board for Biosecurity
OBA	Office of Biotechnology Activities
PNAS	Proceedings of the National Academy
SARS	Severe Acute Respiratory Syndrome

Notes on Transcription

By the middle of Chapter 5, this book makes use of certain transcription conventions that many readers will be unfamiliar with. Steve Clayman and John Heritage's *The News Interview* served as the main source for these conventions, though what they did was itself informed by a long line of work in the field of Conversation Analysis. As this *Biotechnology, Security and the Search for Limits* is not primarily written for those that examine conversation in its finest detail, I only employ certain of the most significant notational conventions in the transcriptions and well as other commonly used transcribing symbols.

Symbol	Example
*** indicates deleted words for purpose of the non-identification of participants	A little piece of my research work, the last line in the paper, I was doing some work on *** toxins, the last line in the paper *** spawned the little bit of work that I'm doing now
Underlining denotes stressed words	And unless you have a policy of open publishing, you,those assumptions are never going to be right. I mean, why just because it's published in the scientific world that it's used for good rather than bad, who's to define what's good or bad, I mean.
Words in (single parentheses) are the author's guess at what was said when the recording was unclear	I know the seriousness with which people tick the ethics box and, and people don't sit down and weigh up the ethical (.) implications of their research, they quickly start ticking (out the boxes) and I sure this do exactly the same.
Words in ((double parentheses)) are the author's description of what happened.	((Group laughter))
Numbers within single parentheses, such as (0.5),	I'd, I'd like to throw the whole thing open by saying(3.0) it would be (1.0)

Continued

Symbol	Example
indicate the estimated time of the pause in seconds. The symbol '(.)' indicates a micro pause of less than 0.2 seconds.	it feels very artificial to pose this question
Pairs of equal signs indicate:	
(a) Either when two individuals spoke with no intervening silence	MD: My = P1: = How do you read it?
(b) When the same individual continued at statement without pause but that statement was broken up overlapping talk from another individual	BR: So this was maybe = P1: [Yeah] BR: = a bit superfluous?
Overlapping talk between two individuals is designated by square brackets. The start of an overlap is indicated by '[' in two adjacent lines. The end of the overlap is indicated by ']' in those separate lines. The ']' might indicate that the overlap ends at the same time or that one speaker carries on after the initial overlap.	BR: ... military issues or [anything] P10: [Well] I think this is a very handy way ...
Where a remark by one individual is inserted into a continuous statement by another, but without any overlap of words, only one set of brackets is used.	BR: So this was maybe = P1: [Yeah] BR: = a bit superfluous?

In addition to these conventions, in the transcribed excerpts 'P' denotes a participant's statement and the number next to that indicates the order in which that participant first spoke in the group discussions. For instance, 'P2' indicates the second participant that spoke in the seminar and any subsequent statement by this person throughout the rest of the discussion is likewise indicated by 'P2'. 'BR' and 'MD' are used to indicate when Brian Rappert and Malcolm Dando were the speakers. In some excerpts, contributions from more than one member of the audience are indicated by 'AUD'.

Introduction: What Should be Done?

In early October 2005, numerous reports appeared in the scientific press and general news media regarding research conducted on the '1918 Spanish Flu'. During 1918 and 1919, this virus was responsible for the deaths of tens of millions of people in a worldwide pandemic. Unlike many strains of influenza that targeted the newborn, elderly, or unhealthy, the 1918 virus killed healthy adults. An estimated one-third of those infected died from breathing complications, contracting pneumonia, or other debilitating conditions. Nearly as quickly as it emerged, the virus vanished. Medical doctors of the time had little understanding of its origins or characteristics.

The attention given to 1918 Spanish Flu in the autumn of 2005 centered on two articles that appeared in prestigious scientific journals. One was the publication in *Nature* of the sequences for the remaining unsequenced parts of the virus's genome.[1] A team lead by Jeff Taubenberger of the US Armed Forces Institute of Pathology had reconstructed the genome by assembling fragments from hospital specimens collected during the pandemic and the remains of victims buried in permafrost. Through an analysis of its genetic makeup, they argued that rather than being a mix of human and avian (bird) flu, the 1918 virus was 'more likely an entirely avian-like virus that adapted to humans.' The second article published in *Science* described the reconstruction of the virus by a research team headed by Terrence Tumpey at the US Centers for Disease Control (also including Jeff Taubenberger).[2] They used the sequence data provided by the Institute of Pathology to physically recreate the virus. Laboratory mice and other specimens were then infected with the 1918 flu strain and genetically modified variants to test the function of certain genes in the transmissibility and virulence of the virus. Through this process a protein on the surface of the 1918 virus was identified as playing a significant role in making it so lethal.

Some reports of the research projects included decidedly positive appraisals. The *New York Times* quoted a British virologist as saying 'This is huge, huge, huge ... It's a huge breakthrough to be able to put a searchlight on a virus that killed 50 million people. I can't think of anything bigger that's happened in virology for many years.'[3] *Nature* ran its own news story in which another virologist was reported as saying 'It's a landmark ... Not only is this the first time this has been done for any ancient pathogen, but it deals with the agent of the most important disease pandemic in human history.'[4]

Both these news reports though (along with many others), included comments of a different kind. Remarks were included suggesting that while some 'scientists have already hailed the work as giving unprecedented insight into the virus', there were also concerns about darker implications.[5] As stated in the *Nature* news report, while some said the benefits were potentially substantial:

> ... others have raised concerns that the dangers of resurrecting the virus are just too great. One biosecurity expert told *Nature* that the risk that the recreated strain might escape is so high, it is almost a certainty. And the publication of the full genome sequence gives any rogue nation or bioterrorist group all the information they need to make their own version of the virus.[6]

A scientist was quoted in the *New York Times* article saying that 'There is a risk verging on inevitability of accidental release of the virus and a risk verging on inevitability of deliberate release.'[7] In a later article for the newspaper, two technology analysts called the decision to publish the flu's genome 'extremely foolish' since it amounted to 'the design of a weapon of mass destruction'.[8] In anticipation of possible concerns and the behest of others (see below), the authors of the article in *Science* included a note which stated: 'This research was done by staff taking antiviral prophylaxis and using stringent biosafety precautions to protect the researchers, the environment, and the public. The fundamental purpose of this work was to provide information critical to protect public health and to develop measures effective against future influenza pandemics.'

The research into the 1918 flu was just one of a number of reasonably high-profile cases since 2001 where concerns had been voiced about the security threats that might stem from work in the life sciences. Some of these have asked whether the research should have been openly communicated or even undertaken at all. Various discussions have taken place in science policy arenas regarding whether some communication restrictions were justified and whether their introduction might have compromised scientific advancement. As in many accounts of the sequencing and recreation of Spanish Flu, the issues at hand in these were widely regarded as highly problematic since the knowledge and techniques generated had the potential for 'dual use': they could aid in the fight against disease or be used to further spread it. Thus the matter of what should have been done raised fundamental questions about the place of science in society.

The proper governance of research deemed 'controversial' was directly addressed in the *New York Times* and *Nature* news reports. As both noted, not only did the 1918 flu research go through conventional safety assessments as well as institutional and peer review, the journals had in place their own security review system. In these the benefits of publication were considered against potential security risks, with the former deemed much more substantial. In addition, the virus reconstruction article in *Science* was scrutinized by the newly established National Scientific Advisory Board (NSABB) at the request of US Health and Human Services Secretary Michael Leavitt. NSABB had been established, in part, to devise criteria and procedures for appraising dual use research. Dr Anthony Fauci of the National Institute of Allergy and Infectious Diseases in the National Institutes of Health said NSABB had 'voted unanimously that the benefits outweighed the risk that it would be used in a nefarious manner.'[9]

Weighing the risks and benefits

The framing of the decision about publishing as one where benefits had to be weighed against risks was widely repeated.[10] A front-page editorial in *Science* by Philip Sharp (MIT) found it 'reassuring that the NSABB was asked to consider these papers before publication and concluded that the scientific benefit of the future use of this information far outweighs the potential risk of misuse. People may be reassured that the system is working, because agencies representing the public, the scientific community, and the publishing journals were involved in the decision.'[11]

The Editor-in-Chief of *Nature* was less enthusiastic about the new NSABB oversight procedures, suggesting the danger that 'government bureaucracies and committees may push to avoid perceived risks, at the potential expense of benefits to public security.' A week after the initial publications, *Science* ran another front-page editorial, this time by its Editor-in-Chief. This second editorial struck a rather different tone than the first by describing some of the 'backstage' maneuverings regarding the decision to publish. Herein, the suggestion to refer the virus reconstruction article to NSABB was characterized as an '11th-hour intervention from the [Health and Human Services] secretary's office'.[12] The previous week's editorial had been finished 'at the last possible moment', as had the added 'authors' note' which was suggested by NSABB to assuage possible concerns. Significant issues were raised by the editor of *Science*, including 'a real question of authority here. Government officials can advise, and should be listened to thoughtfully. But they can't order the nonpublication of a paper just because they consider the findings "sensitive".'[13] In a rather defiant close, the editor finished by writing 'So would I, given our own convictions, the timing, and what we had learned from our consultations with Gerberding, Fauci, and others, have published the paper even if the NSABB had voted otherwise? Absolutely— unless they had it classified.'[14]

Setting aside disputes about institutional authority and timing, appraisals of what should have been done with the 1918 experimental results were inextricably bound with the way in which possible risks and benefits were handled. Central to this was the identified range of germane matters for consideration. Although scientific and general media reports about the studies shared many commonalities, they also differed. While reports in *Nature* and the *New York Times* gave space to concerns about the accidental releases of the reconstructed 1918 virus based on past experience with other laboratory-held pathogens,[15] others did not mention such fears.[16] Most, though not all,[17] news accounts explicitly linked the 1918 influenza research to the high-profile media concerns at the time about the deaths to birds and some humans in Southeast Asia from the H5N1 avian flu virus. Arguably this linkage underscored the possible benefits of the research. Some accounts went so far as to retrospectively claim the reason for the research was to counter bird flu.[18] *Nature* not only published a paper detailing the remaining sequences of the 1918 virus and a related news story, but along with these the sequence results for a major collaborative study of human influenza A viruses. Attesting to the importance of sequencing virus genomes, the latter promised 'to provide a more comprehensive picture of the evolution of influenza viruses and of their pattern of transmission through human and animal populations.'[19]

While some arguments and commentators could be neatly categorized as justifications or criticisms of the publication choices made, other statements were more ambiguous. Presenting seemingly both pros and cons, Ron Atlas of the Center for the Deterrence of Biowarfare and Bioterrorism at the University of Louisville was quoted in *The Guardian* (UK) as stating that the results heightened concerns about a new pandemic coming from bird flu. In addition, '[a]ssuming this is a replicant of the 1918 flu strain, if it got out, it could initiate disease in humans and given the work they've done, one had to say it would be infectious'.[20] Yet, just what overall appraisal Atlas made of the decisions to publish was difficult to gauge. *The Guardian* also stated that despite the beneficial potential of the work mentioned by some, 'other researchers warned yesterday that the virus could escape from the laboratory. "This will raise clear questions among some as to whether they have really created a biological weapon," said Professor Ronald Atlas ....'[21] Here the report itself was arguably ambiguous in that it is not certain from the wording whether Atlas was one of those researchers taken by *The Guardian* as actually warning of dangers or merely pointing out that unidentified others would be concerned.

Off the front page

While the citing of credible experts offering assessments of the merits of the 1918 virus research was central to reports in the scientific and general presses, by moving beyond mainstream accounts, questions can be asked about just who was cited and the bounds of the debate subsequently generated. Although news reports in *Nature* and *Science* described numerous concerns about the research articles they published, these pitched possible safety and misuse risks against likely benefits to human health. Certain commentators challenged this way of thinking, though their assessments were not widely picked up. The Sunshine Project, a watchdog of US activities in this area, maintained that 'We are no safer from a pandemic today than yesterday. In fact, we're in greater danger, not only from influenza; but from the failure of the US to come to grips with and address the threats posed by the research it sponsors, in terms of legislation, ethics, and self-restraint.'[22]

In support of this position, the organization made a variety of points that did not figure in many mainstream accounts: a proposed lack of valid scientific justification for the 1918 flu research projects, the way in which these activities blurred the boundary between internationally permissible defensive and non-permitted offensive biological weapons-related work, the inappropriateness of the massive increase in US research into dangerous pathogens post 9/11, and the past and likely spread of viruses with 1918 genes to many laboratories. Accounts of the motivations for the 1918 Spanish Flu research figured in reports by the Sunshine Project. One of its 2003 publications argued:

> It appears that this work was not triggered by a search for flu treatments, or the search for a new biowarfare agent, but by a rather simple motivation: Taubenberger and his team were just able to do it. In previous experiments they had developed a new technique to analyse DNA in old, preserved tissues and for [sic]now looking for new applications: 'The 1918 flu was by far and away the most interesting thing we could think of', explained Taubenberger [giving] the reason why he started to unravel the secrets of one of most deadliest viruses known to humankind.[23]

The telling of motive here (based on a web profile of Taubenberger) served to reinforce the said distance of the 1918 research from practical health concerns. Taken together, these points provided a counter-narrative to the dominant stories presented elsewhere. This one questioned not just the actions of individual scientists or publishers, but the policies of the US government.

The reference to the 2003 Sunshine Project report points to the pre-history of debate about 1918 flu research prior to the international high-profile attention in October 2005. While many of the widely voiced concerns expressed in 2005 about the 1918 virus research centered on the adequacy of the protections in place to prevent its accidental release and deliberate misuse, the Sunshine Project had drawn attention to a prior question: whether the research should have been undertaken in the first place. Although accepting the need to sequence the 1918 strain, it said: 'there is no valid reason to recreate the virulent virus, as the risks far outweigh the benefits.'[24] Rather than conducting the latter said dangerous research, in 2003 it said funds could be better spent on existing naturally occurring diseases such as malaria and HIV.[25]

In contrast, many of the reports prior to October 2005, concerns about the deliberate misuse of research did not figure into what was reported. Taubenberger and colleagues, for instance, wrote an article for the journal *Scientific American* in January 2005 that recounted various lines of work taking place into the 1918 virus, but without any reference to their potential to facilitate novel bioweapons.[26] In this article, rather than their work being motivated out of mere curiosity, it was said to be motivated by a desire to devise treatments and measures for alleviating any future pandemic outbreaks. A news report in *New Scientist* in 2004 discussed research led by Yoshihiro Kawaoka that entailed inserting some of the genes from the 1918 virus into influenza strains, but again without citing any causes for concern.[27]

The usefulness of research

As mentioned, the Sunshine Project not only criticized the 1918 virus reconstruction research on the basis of concerns about its negative implications but also its said limited utility. Considerations of usefulness figured widely within accounts both during and before October 2005. The utility of the projects in *Nature* and *Science*, however, was a matter alternatively characterized. Some accounts of the research and of experts' appraisals of it were rather categorical and definite. The news article in *Science* said 'researchers have figured out the traits that made the 1918 influenza virus, which killed between 20 million and 50 million people, so virulent.'[28] It went on to state that the team involved in the recreation efforts said 'its work will provide crucial knowledge for heading off the next influenza pandemic, which could be brewing in Asia, where the H5N1 bird flu has killed more than 60 people.'[29] The authors' note, added to the *Science* reconstruction paper at the request of NSABB, drew attention to the practical benefits of the research in stating the 'fundamental purpose of this work was to provide information critical to protect public health and to develop measures effective against future influenza pandemics.' SciDev.Net reported that 'Taubenberger said that studying the 1918 virus should provide a "checklist" of genes that H5N1 would have to acquire to become infectious enough to cause a pandemic.'[30] Such statements suggested the research had already yielded valuable insights that provided concrete justifications for the work.

Other accounts though were more reserved in their claims about the imminent benefits of the research for the state of knowledge or therapeutic interventions.[31] The news story in *Nature* said that future follow-on research would 'hopefully be of use in vaccine and drug design, but so far the work is more about obtaining a basic understanding of the virus than any immediate health benefits.'[32] Rather than stating the work should provide a 'checklist' of genes for studying how bird flu could become a human pandemic, the *New York Times* article reported Taubenberger as contending 'The ultimate goal … is to make a checklist.'[33] A joint statement by the directors of the National Institute of Allergy and Infectious Diseases and Centers for Disease Control and Prevention offered a highly qualified assessment that the 'mysteries of the 1918–1919 influenza pandemic … are finally beginning to be solved'. The directors said the *Science* and *Nature* publications may help predict pandemics and 'may lead to identification of new targets for drugs and vaccines' but more research would be required.[34] While individual publications might not in themselves necessarily lead to significant outcomes, the *New York Times* article stressed the importance of continuing with an unfettered program of research. At this level of the program, it said 'the certain benefits to be obtained by a robust and responsible research agenda aimed at developing the means to detect, prevent and treat these threats far outweigh any theoretical risks associated with such research.'[35]

An October 2005 report by *New Scientist* adopted a fairly mixed position about the usefulness of the research. It questioned a statement made by the leader of the reconstruction of the 1918 virus that 'This work will help us make vaccines and antivirals against pandemic strains' by adding its own comments that 'It is unclear how, as the next pandemic is likely to be a different kind of flu.' Yet immediately following this challenge the article stated: ' "But", says Taubenberger, "the 1918 sequences are already helping in another way: they prove that a bird flu can go pandemic without combining with a human flu, and suggest which mutations it needs." '[36]

Additional queries about the practical utility of the publications were noted elsewhere. While highly supportive of the decision to undertake and publish the research, in his *Science* editorial Philip Sharp noted that the reconstructed virus was based on one full genome sequence. Given variations in the genetic makeup within viruses and uncertainty as to the exact causes of pandemics, he concluded 'there is no certainty that the reconstructed 1918 virus is capable of causing a pandemic'. The claim that the particular virus studied might be significantly dissimilar to that (or those) which caused the brunt of the pandemic performed a number of functions. It at once suggested a reduced utility of the results than that given in many other accounts, justified the need for further research to determine the identity of the reconstructed virus, and reduced the potential harm from the deliberate or accidental misuse of the research.

While most accounts did not raise doubts about the relevancy of the particular genome sequenced for grasping the Spanish Flu pandemic, some gave reason for downplaying concerns about the possible negative consequences of laboratory release. Both the Reuters and *New Scientist* reports included statements by the director of the Centers for Disease Control and Prevention that the human immunity developed from the 1918–19 pandemic meant any release of the virus would limit the risks.[37] In such accounts, arguably a tricky line was walked of both treating the reconstructed flu virus as providing vital clues about the deadliness of viruses while also claiming the actual virus studied was not that dangerous today.

How to decide what should be done?

The case of the 1918 influenza virus has been used to introduce something of the problematic issues often identified as relevant when considering dual-use research. To the question of 'what should be done?', much has been and still might be said. So too might much be said to the question of 'how should we determine what should be done?'.

There are a variety of possible responses to this latter question in light of the contestation suggested above. One would be a call to redouble efforts to establish credible expert assessments regarding the factual matters often presented as central to making appropriate decisions. So in relation to Spanish Flu, this might mean trying to determine whether the research under question actually provided a checklist of genes for assessing the pandemic potential of particular viruses, what general scientific utility it really fulfilled, whether the information given in the scientific publications actually would enable certain nations or groups to construct viable bioweapons, and so on. Through answering such factual questions, authoritative assessments could be made about the advisability of undertaking or publishing the research in question. As indicated, oversight organizations such as NSABB take the devising of criteria for identifying dual-use research risks and benefits as their *raison d'être*. In such a search for 'the facts', any remaining unknowns or uncertainties would threaten to render decision-making problematic.

Related to this approach, one might start with an examination of the adequacy of the general discussions about the research and the formal procedures for making decisions. So in the case of 1918 Spanish Flu, one could ask questions such as: Were the discussions balanced in terms of views presented? Were the scientific experts so central to many accounts and the ultimate decisions taken by *Science, Nature* and NSABB credible and unbiased? Were such expert views sufficient or should input have been sought from others – in other words, who is the 'they' that should partake in determining what should be done? Should concerns which did not gain mainstream attention (such as those of the Sunshine Project) have received greater notice? Might public discussions themselves bring ill-advised prominence to the potential for the malign use of science? Should there have been greater discussion of the implications of the research prior to the October 2005 publications?

Questions can then be asked about such questions: What constitutes a 'balanced' presentation of views? Is this balance, for instance, to be required of individual accounts or in the overall mix of discussion generated? What criteria should be used and what information collected to determine what counts as credible or biased? Who should decide such matters? What roles can those without recognized scientific credentials play in making decisions about issues with highly technical dimensions? To what extent is it possible to assess the implications of research in advance?

Like other concerns voiced in dual-use debates, for many people answers to these sorts of questions would depend on more than just narrow factual scientific data; rather on how science as an activity is conceived. For example, the answers would be required to questions such as: Are there any limits on what should be known or asked? Do the personal motivations of scientists matter in assessing the ultimate merits of work? To what extent must research be directly geared to societal concerns in order to be of societal benefit? How easy is it to convert fundamental research findings in the life sciences into practical applications – benign and malign applications? Is it enough that scientists take care to work in a diligent manner or should they consider possible future implications of that work? If the latter is the case, how should this be done: for instance, openly in the spirit of encouraging robust and widespread social debate or selectively among those with highly specialized expertise?

It is possible to point questioning in a different direction though. Rather than examining the (sometimes competing) appraisals of what should be done or the processes by which these determinations are made, one option is to turn attention to how analysis into these matters is undertaken. For instance, questions can be posed of how the account given above has itself presented a particular and contingent account of what is at stake in discussions about dual-use research. Just as each of the news reports and research articles mentioned above utilized a certain description of events and contexts for consideration, so too did the description given in this chapter. Certain details about the issues under consideration and specific excerpts from articles were utilized to suggest the basic nature of the debate. The adequacy of these claims is a matter open for discussion.

Outline of chapters

Biotechnology, Security and the Limits of Research examines a variety of substantive and methodological issues associated with dual-use research – that is, what actions should be taken in response to concerns about the malign potential of research, how determinations of this are made, and what is at stake in the way inquiry is conducted into these issues. Each of these raises challenges and choices; challenges and choices in determining what sort of biological or social research should be done as well as in what questions should be asked as part of that research. In this manner, this book reflects on the place of life science research in contributing to and responding to public concerns as I reflect on how I undertake research into these issues. One of its central aims is to further an appreciation of the place of social and life science research in contemporary public issues.

Chapter 1 begins this by reviewing recent US-centered discussions and initiatives regarding the security implications of bioscience and biomedical research. Far reaching, questions have been posed in these actions. This chapter aims to provide an overview of them. In doing so it approaches assessments of the 'dual-use' potential of research through themes in the 'social problems' literature. This is to say that this chapter attends to the contingent, collective, and changing processes whereby dangers are defined, proposals for action are tabled by certain individuals, and choices are made about what needs to be done. From an analytical point of view, a key issue raised through this orientation is whether social analysts of problems should propose definite assessments about their nature or confine themselves to examining how such determinations are reached by others. Considering the differences and inter-relation between such 'objectivist' and 'constructivist' tendencies will provide an initial introduction for recurring attention in this book to the choices made in undertaking social research.

As has been argued in this introductory chapter and as will be contended in Chapter 1, assessments of the nature of dual-use research and what should be done in response often turn on assumptions about what scientific research is like as a practice. The second chapter, 'Discussing Science' reviews the characteristics often attributed to scientific research in dual-use policy discussions. Specifically, attention is given to the role of openness and freedom in the pursuit in science and how evidence is marshaled in validating knowledge. The chapter then turns to examining what social analysts of science have made of these said characteristics. The contrasts in social analyses of science – in terms of their methods, evidence, audience, and purpose – are used to underscore the need to attend to what is taken as given and what is questioned when dual-use issues are examined.

Chapter 3 continues, asking 'What should be done?' both for the substantive issues associated with dual-use life science research as well as social research into this topic. It initially does so by reviewing the importance attached to education and raising awareness. Tensions and questions associated with the nature of any educational program for scientists are surveyed to underscore the importance of facilitating processes of *learning* in research into this topic: learning about the threats associated with bioweapons, learning about what makes for a prudent responsive measure, and learning about how to investigate it. On the back of these considerations, the argument is made for conducting social research in a problem-orientated fashion that strives for further mutual learning. The philosophical tradition of pragmatism – as exemplified by the works of John Dewey – is forwarded as a starting basis for thinking about forms of inquiry that aim to fit this goal.

Chapter 4 then moves to elaborate a specific line of empirical research in dual-use issues undertaken by the author and Malcolm Dando. In an effort to engage scientists in discussions about what should be done, we undertook a modified form of focus group research in the United Kingdom. This chapter recounts the substantive responses to questions while also attending to the methodological matters of how questioning was undertaken. As will be detailed, in an effort to pose the question of what kind of questioning was required, the basic orientation adopted *within* individual seminars and in the transition *between* seminars was one of making explicit the data, assumptions, and inferences underlying responses in order to publicly challenge those in aid of learning.

Through a detailed analysis of empirical material and exchanges conducted after the seminars, the remaining chapters of Part II extend the parallel attention in previous chapters to what is known and how it is known. This is done through the examination of key tensions related to both substantive choices about what should be done about dual-use research as well interactional choices in how conversation about what should be done are organized. Those tensions include the relation between openness and closure, neutrality and bias as well as expertise and equality. In doing so, the chapters in Part II seek to maintain a productive, but problematic relation between substantive responses and interactional concerns. This will be done through moving away from the general recounting of methods and strategy as done in Chapter 4. Instead, much more attention is given to the actions of seminar participants and moderators in particular exchanges. As will be contended in Chapter 5, doing so also justifies rethinking how information about exchanges should be reported.

Part III then closes the argument of the book with a summary that places particular emphasis on what has been learnt and what can be learnt from the inquiry described in this book.

Part I

Scientific Research and Social Inquiry

1
The Dilemmas of Dual-use Research

The 'Introduction' recounted some of the debates about 1918 Spanish Flu research to suggest something of the dynamics of discussions about the dual-use applications of the life sciences. Questions were posed not only about the choices and challenges in conducting biological research but also in examining dual-use issues in the first place. With regard to both, questions can be posed about what needs to be done and what is at stake in the way research is undertaken.

This chapter continues with this examination of policy debates while also reflecting on how this is done. In doing so it seeks to question what constitutes a proper understanding of 'the problem' of dual-use research. This is accomplished, in part, by treating dual-use research as a social problem; which is to say attention is given to the contingent processes whereby dangers are identified, proposals are tabled, and choices are made about what needs to be done.[1] While developments in science and engineering have long provided the basis for more sophisticated weaponry, in recent years the attention devoted to the life sciences and bioweapons has increased substantially. That attention has inevitably focused on certain issues in a certain manner. As suggested from the case of Spanish Flu, just how the link between 'research' and 'threats' ought to be conceived can be a matter of contestation. The changing, sometimes multiply envisaged, dangers identified are a topic for scrutiny in this chapter. In this way it sets the preliminary parameters for an understanding of the potential of research that will be drawn on and also subjected to scrutiny in later chapters.

Scrutiny is also given to what this social problem's orientation itself implies for the understanding given to 'the problem' of dual use. With a notional focus on how concerns emerge and get defined, in this chapter I do not intent to offer a definitive account of the scale of the dual-use potential of the life sciences. Instead, I choose to ask how individuals and organizations have sought to forward specific definitions and characterizations of this. As a principle of methodology, such a move assumes that the definition of a problem is open to multiple, perhaps fundamentally opposed, interpretations. As societal concerns about issues such as homelessness, poverty, hooliganism, and road rage 'may emerge, transmute, descend, disappear, or reappear, independent of any change in actual conditions',[2] so too can the understanding given to the implications of science. Therefore, talk of the real problems and possibilities posed from life science research is largely eschewed in favor of considering how certain claims are justified and their implications for determinations of what needs to be done. As will be suggested, however, following through in this orientation raises a number of important questions about the adequacy of the analysis obtained, what it implies about the responsibilities of social researchers, and the relation between such analysts and public issues.

A statement of the problem of 'dual-use research'

One of the highest profile and arguably most influential statements about the weapons potential of life science research was the US National Research Council (of the National Academies) report *Biotechnology Research in an Age of Terrorism*. Published in October of 2003, the report was billed as the first time that the National Academies directly examined the national security issues associated with the life sciences. It was a product of the Council's recently established Committee on Research Standards and Practices to Prevent the Destructive Application of Biotechnology. The Committee held a number of meetings from April 2002 under the remit to offer advice to the federal government regarding what might be done about the threats of bioterrorism and biowarfare from fundamental research. The Committee was among a number of other groups and events initiated by the National Academies to consider science and security issues post-9/11.[3]

As stated in *Biotechnology Research in an Age of Terrorism*, the problem that needed addressing was 'the intentional use of biotechnology for destructive purposes'.[4] Consideration of the security implications associated with the life sciences was justified by recent developments including, 'the discovery of nations with clandestine research programs dedicated to the creation of biological weapons, the anthrax attacks of 2001, the rapid pace of progress in biotechnology, and the accessibility of these new technologies by the Internet.'[5] The Committee did not equally reflect on every aspect of intentional destructive use of biotechnology. Citing the legislation already introduced post-9/11 regarding the possible diversion of dangerous agents from laboratories and the screening of lab personnel, focus was given to how the technologies, methods, and knowledge generated in advanced research might facilitate the creation of bioweapons.

In the report, the nature of science as an activity was central to understanding the nature of the dual-use problem. Professor Gerald Fink of the Whitehead Institute for Biomedical Research chaired the Committee and succinctly summarized many of the points made when he argued:

> [A]lmost all biotechnology in the service of human health can be subverted for misuse by hostile individuals or nations. The major vehicles of bioterrorism, at least in the near term, are likely to be based on materials and techniques that are available throughout the world and are easily acquired. Most importantly, a critical element of our defense against bioterrorism is the accelerated development of biotechnology to advance our ability to detect and cure disease. Since the development of biotechnology is facilitated by the sharing of ideas and materials, open communication offers the best security against bioterrorism. The tension between the spread of technologies that protect us and the spread of technologies that threaten us is the crux of the dilemma.[6]

Thus, concerns about destructive applications of biology posed a vexing dilemma, since the promise of biotechnology went hand in hand with its darker side.

The comments made by Fink were expanded on in the report and marshaled to justify the recommendations reached. So the imperative was frequently forwarded to not jeopardize the 'great achievements of molecular biology and genetics over the last 50 years'[7] because of the potential misuse of science. The advancement of science was itself predicated on the 'norm of open communication'.[8] As stated, '[t]he rapid advance of scientific knowledge and applications owes much to a research culture in which knowledge and biological materials are shared among scientists and people move freely between universities, government agencies, and private industry.'[9] In part because of this free movement, science was also very much an international activity. So, '[w]ithout international consensus and consistent guidelines for overseeing research in advanced biotechnology, limitations on certain types of research in the United States would only impede the progress of biomedical research here and undermine our own national interests.'[10]

While *Biotechnology Research in an Age of Terrorism* stated that the potential for the deliberate destructive use of biotechnology was not new to the beginning of the twenty-first century, a variety of reasons made concern about any threats especially salient today. This included three recent experiments discussed in some detail illustrating the malign possibilities enabled by advanced biology. One, the insertion of the interleukin-4 gene (IL-4) into the mousepox virus by Australian researchers in early 2001 to find an infectious contraceptive for reducing animal populations.[11] With the high mortality rates achieved for immunized and non-immunized mice because of the over-expressed IL-4, it was feared this experiment suggested a technique for enhancing the lethality of other pox viruses (such as smallpox). Second, the 2002 announcement of the successful artificial chemical synthesis of poliovirus that signaled a way to create other viruses from scratch.[12] Third, the comparison of a type of smallpox and its vaccine published in 2002 that proposed a means of making the vaccine more lethal.[13]

Just as the destructive potential of biotechnology was not portrayed as novel to the twenty-first century, so too were national security concerns about research not portrayed as new either. The Committee compared the fields of nuclear science and cryptography with that of the life sciences and concluded that 'controls on information flows in the life sciences will face obstacles rather different ...'.[14] Factors such as the size of the life science community, the number of publications produced per year, and the lack of previous engagement with security concerns meant that models for implementing security measures derived from these other fields would not work.

Instead of proposing oversight procedures modeled on practices elsewhere, the Committee recommended extending many of the community self-governance mechanisms already in place in the life sciences. One of the recommendations, for instance, called for the initiation of a system of pre-project review of so-called 'experiments of concern' by extending existing biosafety and recombinant DNA review procedures. Box 1.1 lists the main recommendations advocated. A theme stressed throughout the report was the importance of extending any initiatives beyond the US.

While setting out an agenda for action, the report left many difficult issues unresolved. The Committee both argued for the need to introduce new measures that might restrict research and for the need to ensure that such measures did not 'impinge upon the ability of the life science community to continue its role of contributing to the betterments of life and

Box 1.1 Recommendations from *Biotechnology Research in an Age of Terrorism*

Recommendation 1: Educating the Scientific Community

We recommend that national and international professional societies and related organizations and institutions create programs to educate scientists about the nature of dual-use dilemma in biotechnology and their responsibilities to mitigate its risks.

Recommendation 2: Review of Plans for Experiments

We recommend the Department of Health and Human Services (DHHS) augment the already established systems for the review of experiments involving recombinant DNA conducted by the National Institutes of Health to create a review system for seven classes of experiments (the Experiments of Concern) involving microbial agents that raise concerns about their potential for misuse.

Recommendation 3: Review at the Publication Stage

We recommend relying on self-governance by scientists and scientific journals to review publications for their potential national security risks.

Recommendation 4: Creation of a National Science Advisory Board for Biodefense

We recommend that the Department of Health and Human Services create a National Science Advisory Board for Biodefense (NSABB) to provide advice, guidance, and leadership for the system of review and oversight we are proposing.

Recommendation 5: Additional Elements for Protection Against Misuse

We recommend that the federal government rely on the implementation of current legislation and regulation, with periodic review by the NSABB, to provide protection of biological material and supervision of personnel working with these materials.

Recommendation 6: A Role for the Life Sciences in Efforts to Prevent Bioterrorism and Biowarfare

We recommend that the national security and law enforcement communities develop new channels of sustained communication with the life sciences community about how to mitigate the risks of bioterrorism.

Recommendation 7: Harmonized International Oversight

We recommend that the international policy-making and scientific communities create an International Forum on Biosecurity to develop and promote harmonized national, regional, and international measures that will provide a counterpart to the system we recommend for the United States.

improving defenses against biological threats'.[15] This overall tricky desire to curb without impeding was repeated in relation to specific actions. So while, on the one hand, it was said that '[t]o limit the information available in the methods section of journals articles would violate the norm that all experimental results should be open to challenge by others', on the other hand, it was noted that 'not to do so is potentially to provide important information to biowarfare programs in other countries or to terrorist groups.'[16] Arguably important to determining what ought to be done by the way of dissemination controls were assessments about the feasibility of limiting the spread of research. While the report warned about the dangers of controls for the progress of science, it also stated that there was an 'inevitable diffusion of knowledge and capabilities' to states and non-state groups.[17]

From the perspective of public policy, *Biotechnology Research in an Age of Terrorism* could also be noted for what it did not say. No attempt, for instance, was made to detail the magnitude and severity of dual-use concerns or propose how the threats from this compared with other terrorist or natural threats.

A brief recent history of 'dual-use' life science research

With its explicit focus on the possible restrictions of civilian fundamental life science research, the agenda for *Biotechnology Research in an Age of Terrorism* differed in noteworthy respects from the vast majority of *pre*-9/11 policy analyses. While concerns about the potential for genetic techniques to aid the development of bioweapons have been voiced since the development of such techniques, this has not translated into such plans for national oversight procedures.[18] So although a 1997 US Department of Defense report identified the following trends as influencing the likelihood that infectious agents would be used as bioweapons:

- Genetically engineered vectors in the form of modified infectious organisms will be increasingly employed as tools in medicine and the techniques will become more widely available.
- Strides will be made in the understanding of infectious disease mechanisms and in microbial genetics that are responsible for disease processes.
- An increased understanding of the human immune system function and disease mechanisms will shed light on the circumstances that cause individual susceptibility to infectious disease.[19]

Its recommendations for action were limited to fairly long-established concerns about strengthening counter-proliferation measures. Likewise, while events such as the First US National Symposium on Medical and Public Health Response to Bioterrorism in 1999 and the Second Symposium in 2000 signaled the increasing policy attention to bioterrorism in the late-1990s, the recommendations made by contributors dealt with matters such as devising threat scenarios, developing early warning sensors, co-ordinating public health responses to attacks, and preventing individuals from former biological weapons programs selling their skills to the highest bidder.[20] Even up to the days before 9/11, analyses of the how the life sciences might aid the design of new bioweapons shied away from considering controls on research findings.[21]

In examining the contingent processes whereby threats from research are identified as a problem that needs addressing, this section outlines key emerging issues between 9/11 and the publication of the *Biotechnology Research in an Age of Terrorism* report.

As noted previously, much of the initial policy attention in the United States after 9/11 and the anthrax letter attacks in 2001 centered on strengthening the physical containment of pathogens and the vetting of personnel working with dangerous 'select' agents. The 2001 US *PATRIOT Act* and the later *Public Health Security and Bioterrorism Preparedness and Response Act of 2002* brought in a variety of enhanced controls on the registration, transfer, storage, and use of recognized dangerous agents.[22]

Yet from the autumn of 2001, articles in the scientific and general press raised questions beyond these more traditional matters.[23] For instance, in October, *Nature* ran a feature article called 'The Bugs of War', outlining a wide ranging set of possibilities for how knowledge and techniques in microbiology and genetic engineering could aid in the development of sophisticated bioweapons. This included genetically manipulating viruses to increase their virulence and survivability, hybridizing viral strains, or introducing antibiotic resistance into bacteria.[24] The case of the insertion of the IL-4 gene into the mousepox virus was the central example of how otherwise benign research might have other applications. Technical possibilities for detecting and countering biothreats were discussed, from which it was concluded that 'the techniques that could produce bioweapons are also being deployed to set up countermeasures against them. This neatly illustrates the point that legitimate and malevolent applications of biology are merely two sides of the same coin.'[25] The article ended with an appeal from Matthew Meselson at Harvard University that it was 'time for biologists to begin asking what means we have to keep the technology from being used in subverted ways'.[26]

A month later *Nature* published another feature article, which included a much sterner call to action.[27] This was based on an interview with George Poste, former head of research at the pharmaceutical firm SmithKline Beecham, and then chair of a US Department of Defense task force on bioterrorism. Poste lamented about how security sensitive research entered the public domain. The IL-4 mousepox experiment likewise figured centrally as an example of this spread. Instead of merely calling for more discussion, he said that biology must 'lose its innocence',[28] to prevent unduly draconian legislation be introduced from elsewhere. Possible, largely self-policed, controls included greater classification of findings, the inclusion of questions about the possible malign application of research on grant submission forms, the vetting of scientific articles, and the limiting of access to certain sequence information. Scientists such as Anthony Fauci from the NIAID were quoted as expressing various reservations about such measures while arguing for greater awareness and debate about the malign applications of science. Poste's call was repeated in an article for the *Financial Times* in which he wrote that 'The issue is not to retreat to "forbidden knowledge" in which areas of research are prohibited. Rather it is about defining "constrained knowledge" in which freedom of research is not impeded but public access to certain categories of research data is restricted.'[29]

Along with such stories came more in-depth analyses.[30] In an extended article in late-2001 for *Critical Reviews in Microbiology*, Epstein examined the possible contribution of civilian research to devising novel bioweapons. In it he offered the category of 'contentious research' to refer to 'fundamental biological or biomedical investigations that produce organisms or knowledge that could have immediate weapons implications, and therefore raise questions concerning whether and how that research ought to be conducted and disseminated.'[31] Epstein called for the scientific community to initiate a process of dialogue with the national security community on such research if it did 'not want to operate under a Congressionally mandated governance regime'.[32] The Asilomar Conference of 1975 conducted by biologists that eventually led to the creation of the NIH Recombinant Advisory Committee and institutional biosafety review procedures was presented as a model with some promise.

Subsequent analyses by academics and US governmental agencies offered appraisals of the threats posed by developments in the life sciences and some gave explicit attention to just what needed to be done.[33] For instance, by way of suggesting guidelines for restricting publications, Zilinskas and Tucker identified a number of lines of contemporary research that posed dual-use dilemmas (see Box 1.2).[34] They called for any system of publication review to have the 'support of the international scientific community, which must perceive that the security benefits of restricting open publication outweigh the possible costs to science'.[35] Carlson argued that the unstoppable exponential increase in the sophistication and the

Box 1.2 Research that poses dual-use dilemmas (Zilinskas and Tucker)

- Sequencing the genomes of human pathogens.
- Construction of 'fusion toxins' derived from two distinct toxins.
- Genetic engineering of a Bacillus anthracis strain containing inserted toxin genes (for example, for cereolysine AB).
- The finding by Australian researchers that inserting the gene for interleukin-4 into the genome of ectromelia (mousepox) virus significantly enhances viral virulence and vaccine resistance in mice.
- Development of 'stealth' viruses that evade the human immune system.
- Publication of the molecular details of two virulent strains of influenza, the 1997 Hong Kong Flu and the 1918 Spanish Flu, the second of which killed 20 million to 40 million people worldwide.
- Generation of influenza A virus from cloned DNA segments.
- Genetic engineering of the tobacco plant to produce subunits of cholera toxin.
- Studies of viral proteins that are similar to mammalian proteins, as tools to probe their function.
- Aerosol spray drug-delivery systems.
- Synthesis of infectious poliovirus by assembling custom DNA strands ordered from a commercial biotechnology company.

gradual proliferation of biotechnologies would considerably ease the developing of bioweapons in the future. But, he maintained, limits and regulations were not the answer. Instead, what was needed was openness to speedy countermeasures.[36]

As mentioned above, three examples of research figured prominently in many analyses in 2001–03: the IL-4 mousepox experiment, the comparison of smallpox and its vaccine, and the artificial chemical synthesis of poliovirus.[37] With these came attention to questions of whether certain research really had the potential to aid in the development of new weaponry and thus whether it should not have been published or conducted. The fear that the poliovirus experiment (published in *Science*) might provide a technique for enabling the creation of other deadly viruses reportedly led eight members of Congress to put down a resolution to congressional committees about the dangers of the open publication of such research.[38] Yet, its importance was questioned elsewhere, including a letter to *Science* by Steven Block of Stanford University. This claimed the work amounted 'to little more than a stunt'[39] since the possibility of performing such a reconstruction was long understood. Against what was presented as the 'extraordinary press coverage' generated, he argued that techniques for synthetically creating poliovirus were unlikely to pose any security threats in the near term. This was the case not least because of the ease of gathering natural samples of pathogenic viruses and the complexity of recreating smallpox where such a natural gathering was not feasible. The editor of *Science* responded by commenting that Block's letter neatly summarized 'a number of reasons why the national security concerns [in relation to this experiment] are not worth serious consideration; we couldn't have put it better ourselves, and we are grateful for this clarification from a bona fide expert on biowarfare.'[40] Yet the editor countered Block's claims about the non-importance of the research by arguing it proved a principle. Similar concerns about the novelty, worth, and (by implication) the security threats associated with the IL-4 mousepox experiment had also been raised at the time.[41]

During 2002 and early 2003 initial steps were taken to translate the general concern about the destructive application of research findings into policies and actions. The Homeland Security Act of 2002 included provisions requiring the federal government to 'identify and safeguard homeland security information that is sensitive but unclassified'.[42] The Act, though, did not define what the category of 'sensitive but unclassified' entailed, what 'identify' or 'safeguard' meant, or exactly how concerns about homeland security should be reconciled with the access to information. A concern repeatedly voiced about the Act was that this category might be applied to scientific research.[43] In 2002 the US Department of Agriculture requested that the National Academy of Sciences restrict access to a report it commissioned (titled *Countering Biological Terrorism*) out of worries about the potential utility of the information included to terrorists.[44] In early 2003, the NAS convened an informal group of 32 largely American-based journal editors, including those representing the journals of the American Society for Microbiology (ASM). This group agreed voluntary guidelines for reviewing, modifying, and if necessary rejecting research articles where 'the potential harm of publication outweighs the potential societal benefits.'[45]

Science under jeopardy?

Throughout these early discussions post-9/11, fears were frequently raised that any security restrictions or oversight measures might jeopardize scientific practice.[46] One initiative that generated particular concern was a proposal (later withdrawn) by the Department of Defense in early 2002 to make it a legal requirement that its funded researchers obtained authorization to disclose their results. In an article for *Science,* Abigail Salyers (then president of the ASM, the world's largest life sciences organization) warned that while security concerns had to be taken seriously it was crucial not to overreact because:

> Censorship of scientific communication would provide a false sense of protection. For example, deleting methods sections from scientific publications, with the rationale that a terrorist could benefit from knowing the methodology, would certainly compromise our ability to replicate results, one of the cornerstones of scientific research. Scientific colleagues' scrutiny and replication of research studies reduces the likelihood of errors that can misdirect scientific activities ...
>
> The best protection against the possibility of future bioterrorism incidents is the unfettered ability of our scientific community to collaborate openly and move forward rapidly in the conduct of scientific research. Timely communication of new knowledge and technological innovation accelerates the rate of scientific progress.[47]

While Salyers stated that scientists appreciated the open exchange of information, 'the public may not necessarily be convinced that scientists can be trusted to this extent. There remains an undercurrent of public discomfort with what is seen by some, however wrongly, as freedom without responsibility.'[48] Elsewhere, concern about replication was said to underline a reluctance by the ASM to allow researchers to remove 'sensitive' elements from research articles since this would 'alter the fundamental tenets of science by eliminating reproducibility of scientific research and undermining the peer-review process for evaluating scientific merit'.[49] The then president of the National Academy of Science likewise argued that harsh responsive measures might well 'severely hamper the US research enterprise and *decrease* national security'.[50] Similar sentiments were even echoed by government agencies. So John Marburger, Director of the US Office of Science and Technology Policy, commented that:

> Science is inherently a social activity. It thrives only in an environment where ideas can be freely exchanged, criticized, and interpreted by others. For a nation that would lead in science, national security includes securing the freedom to engage in open scientific discourse. Science can never be successfully dictated by a science czar, or conducted by a closed elite. Where the marketplace of ideas is regulated, the quality of thought diminishes, and science suffers.[51]

A workshop held in 2003 by the NRC convened Committee on Genomics Databases for Bioterrorism Threat Agents reiterated such basic sentiments about the importance of a marketplace of ideas when it concluded it was vital to ensure 'rapid, unrestricted public access' to genomic databases.[52]

The emerging sense of the 'dual-use' problem

It was largely through such expert informed publications and presentations that an outline was given of the problem of the destructive use of life science research; a potential which ultimately became widely labeled as one of 'dual use'. While no commentators suggested that this potential was new to the start of the twenty-first century, various scientific and security developments made it a much more salient problem.

Whatever the broad similarities in the basic identification of a problem within US-centered science policy discussions though, recommendations for what needed doing turned on estimates of the threats posed from bioweapons. Assessments of this differed. Post-9/11 and the anthrax attacks; much attention was placed on the use of agents by terrorist groups. For some, the limited number of bioterrorist attacks in the past and the difficulties experienced by even well-funded groups interested in using naturally occurring pathogens (for instance, the Japanese Aum Shinrikyo cult) indicated a low likelihood of mass casualty attacks. Following from this, the possibility that such groups could make use of advanced life science research was even more remote.[53] Yet, others pointed to the mass-distribution rather than mass-casualty potential of any attacks or the possibility of state-sponsored terrorism.[54] Time frames and the breadth of technological developments taken into consideration were important variables in assessments. Concerns were expressed that converging technological developments in the life sciences and elsewhere in the coming decades would enable many states, terrorist groups, or even individual sociopaths to produce sophisticated bioweapons.[55]

In summary, there was widespread agreement about the need for greater attention to the destructive potential of research, but one where there were different assessments of the nature of the threat posed. So, too, was there widespread agreement about the need not to jeopardize the said numerous benefits of research but also disagreement about how this could be achieved. For some, the introduction of any new oversight measures or restrictions of research findings threatened to be the thin of edge of a wedge; one which might parallel the damaging biosecurity controls already introduced on laboratory personnel and international students.[56] For others, including key officials, the real danger lay in the scientific community not taking up these issues as this would require the government to do so.[57]

However, it was widely argued that those from the scientific community should be central to discussions about what to do about dual-use results and techniques. Even those identified as leading figures calling for action, such as George Poste, largely confined responses to activities scientists would initiate and police.[58] The process of devising the report *Biotechnology Research in an Age of Terrorism* and its conclusions embodied this spirit of centering action within the research community. Yet, to just what extent and how other communities should contribute to defining the problem and possible remedies was a matter of expressed uncertainty and unease.[59]

That science was regarded as crucial to countering the threats from bioattacks in government policy was illustrated in the substantial increase in biodefense funding in the United States after 2001.[60] While in 2001 all aspects of US civilian biodefense funding totaled $417 million, by 2004 that had risen to $7650 million. Research on infectious disease lead by the National Institute of Allergy and Infectious Diseases (NIAID) at the National Institutes of Health was a major pillar of that funding response. Its budget for biodefence had gone from $53 million in 2001 to $1629 million in 2004.[61] Funding was focused on 'Category A' traditional agents (e.g., anthrax, smallpox, tularemia, and plague) and broken down along funding streams for therapeutics, diagnostics, host response, vaccines, basic biological mechanisms and building expertise resources. Yet, owing to the 'dual-use' potential of research – as developed in the Introduction regarding Spanish Flu – active debates took place then, and since, regarding whether such research helped alleviate or exacerbated threats.[62]

Post-Biotechnology Research in an Age of Terrorism *developments*

In systematically covering a range of dual-use research from the position of an elite scientific organization, the 2003 National Research Council of the National Academies *Biotechnology Research in an Age of Terrorism* report represented an important milestone in discussions about what needed to be done; one which helped set the US federal policy agenda. In March 2004, for instance, the Secretary of the Department of Health and Human Services announced the formation of the National Science Advisory Board for Biosecurity (NSABB) to provide advice on oversight strategies, guidelines and education regarding the handling of federally conducted or supported dual-use biological research.[63] That included criteria for identifying and evaluating the risks and benefits with dual-use biological research for local Institutional Biosafety Committees and the NSABB at a national level for highly problematic cases.

Questions about potential dual-use research and concerns about how best to take action continued to be raised in response to developing events. The story with the three dual-use experiments at the center of many analyses in 2001 and 2002 did not end then. In late 2003, Craig Venter and others at the Institute for Biological Energy Alternatives published in the *Proceedings of the National Academy* a methodology for assembling segments of DNA which enabled them to synthesize a bacteriophage virus in 14 days. The techniques employed meant viruses could be assembled much more rapidly than had been the case for the reconstruction of polio.[64] Working under a biodefense NIAID grant, in late 2003, a group of researchers at St Louis University modified the introduction of the IL-4 gene into mousepox to make the virus 100 percent lethal against vaccinated mice and those treated with anti-virals. They then introduced it into cowpox.[65]

Other fairly high profile individual experiments or publications attracted attention. In 2005, the *Proceedings of the National Academy* published an article suggesting how terrorists could contaminate the US milk supply with probable mass-casualty effects by introducing botulism toxin within it.[66] The lead author also published an opinion editorial in the *New York Times* describing its findings in more accessible terms.[67]

In these and other such developments, questions were raised about the scientific importance and security implications of the work conducted.[68] Just what was asked varied from case to case, but often included questions such as: How easy was it to turn such findings into the production of bioweapons? Were the assumptions underlying fears of the malign use of research well founded? How likely was it that the results obtained in one area (for instance, the effects of IL-4 on cowpox) hold for another (for instance, smallpox)? Was the research conducted necessary given what was already known? What preventive or therapeutic benefits might it enable?

While certain documented developments received a fair amount of coverage in popular and semi-popular media outlets, others did not. In a presentation convened as part of the 'International Forum on Biosecurity' recommended in the *Biotechnology Research in an Age of Terrorism* report, the editor of *Nature* identified several non-prominent manuscripts submitted since 2003 to *Nature* and *Science* that initiated biosecurity reviews by the journals. This included ones giving the sequences for anthrax and SARS, another detailing the structure of anthrax toxin receptor, and one describing how to use microchips to synthesize complex genomes.[69]

The place and priority accorded to biodefense was a topic that generated significant discussion in science policy circles post-2003 in the United States. In 2004 the presidents of the National Academy of Sciences and the Institute of Medicine called for mobilizing a much wider range of scientists under funding programs than the said hitherto preoccupation with virologists and microbiologists.[70] Echoing an appraisal reached elsewhere,[71] Representative Jim Turner of the House Select Committee on Homeland Security argued that the rapid development of biotechnology meant that future threats would increasingly come from engineered agents. Because of this, biodefense funding had to significantly expand beyond traditional bioagents.[72] Others, though, argued that the emphasis placed on biodefense had already distorted the priorities of publicly funded life science research, created dangers regarding the accidental or deliberate release of pathogens, and threatened to blur beyond recognition the line of internationally permissible defensive work.[73]

Social analysis as evaluating responses to the dual-use dilemma

The previous sections gave a broad outline of emerging discussions centered in the United States regarding the problems of what has commonly become called 'dual-use life science research'. In doing so, a sense has been given of how the destructive applications of results and techniques became defined as a problem that needed and could be addressed.

In investigating this topic, questions can likewise be posed about the choices made in what kind of analysis is undertaken.

For instance, instead of just noting the manner in which the varying definitions and solutions proposed provided a particular understanding of what was at stake, I could have explicitly evaluated the claims forwarded. One way this could have been done would have been by assessing the widespread approach advocated for deciding what to do: this being that existing scientific oversight mechanisms (such as peer review and institutional safety boards) should identify activities of concern, weigh the risks (or costs) and benefits of individual proposals or publications, and then make any necessary responses on that basis.

The risk–benefit framing has been pervasive in biosecurity discussions. Both the Introduction and Chapter 1 indicated a number of reports and review procedures (e.g., NSABB, journal reviews) that have pitched the oversight of research in such terms. In addition, the World

Health Organization, the American Medical Association, and various UK funding agencies have adopted a similar risk analysis framework.[74] It is through such rationalistic assessment procedures that the complicated issues surrounding what needs to be done are intended to be made manageable.

It is possible to question the adequacy of this way of thinking. For instance, in 2004 I delivered two related papers to workshops; one on global security held by the Italian National Science Academy and another on terrorism sponsored by NATO.[75] A variety of theoretical and empirical points were made to cast doubt on the suitability of the logic of weighing risks and benefits in the manner commonly suggested. While advances in knowledge and possible applications from research might be identified, substantiating the perils that might follow and the security benefits of constraints would be much more challenging since little appreciation existed about how research would be taken forward for malevolent ends. In addition, when appraisals would be made of whether misuse or counter-misuse ends were most served by a particular piece of research, this would almost certainly come down on the side of those countering threats since they would have vastly disproportional expertise and resources. Related to this, given the general emphasis placed on staying ahead of threats through innovation, the identification of 'contentious' knowledge was unlikely to be regarded as needing to be concealed or limited. It was just such knowledge that had to be circulated and pursued if one wanted to stay ahead through innovation. I argued that these conclusions were supported by experience up till that time with those journals that had initiated voluntary security review procedures – specifically in the lack of any publication refusals and the modification of only two of the tens of thousands of manuscripts submitted.

As such, the fear expressed in some quarters that oversight measures would impose significant limits on experimentation seemed doubtful. As I contended, for those who regard the biosecurity preoccupations as motivated out of political hype rather than scientific reality or for those who otherwise regard the threat as relatively minor, this might have been regarded as no bad thing. The oversight measures undertaken could be positively regarded as a way of being seen to be doing something without needlessly impeding vital research. Yet, even within this way of thinking, whether pursuing this path made for good policy was more questionable given the efforts that would be required.

Another area of concern was that the framing of weighing the risks and benefits of individual pieces of research risked marginalizing concerns about the cumulative developments in the life sciences. So rather than centering security attention on isolated findings, it made more sense to consider major paths of research as a whole and, in particular, what this meant for the proliferation of enabling capabilities. As such, a more pertinent question to be asked than 'Is this finding dangerous?' is 'What is being made routine?' Likewise, rather than focusing on questions such as 'Should this particular experiment go ahead?', it made sense to ask 'What direction of research should be funded?' An exception to the otherwise absence of such system thinking has been in the debate initiated since by leading scientists in synthetic biology about the field's direction and implications.[76]

As an additional concern, I questioned whether the managerialist preoccupation with assessing research was sufficiently visionary. With the moral anguish that followed the use of the atomic bomb in Hiroshima and Nagasaki during World War II, for instance, physicists posed demanding questions about the role of their science in securing international security. Leading physicists such as Niels Bohr, Robert Oppenheimer (at least for a time) and others sought to employ the recognized possibilities afforded by science as a means of pressing for new forms of international relations that would establish a more peaceful world. As was argued, the threat posed by nuclear weapons could only be adequately addressed through international openness in matters of security coupled with a major rethink in the way arms were controlled. As the handmaidens for this nuclear age, some physicists suggested that their community had a special responsibility to be at the forefront of promoting a new political universalism. Just what vision the life science community had for the future of international relations at the start of the twenty-first-century context was not at all clear. As I argued, while far-reaching proposals had been offered by defense analysts at the Center for International and Security Studies regarding the oversight of research,[77] voices within the life science community challenging commitments, advocating new forms of international transparency and co-operation, or promoting ambitious prohibition regimes had been rather muted.[78]

In the study of social problems, some have argued that policies and methods to do with intractable problems that defy easy redress (such as poverty, gambling, and prostitution) sometimes 'have other aims that are equally or even more important than the solution of control of a given problem – namely, the need to demonstrate that the problem is being addressed. Such demonstration shows a commitment to the maintenance of social order and serves as a public assertion, or re-assertion, of dominant values and interests, and the legitimation of such values and interests.'[79] Given the various points made about dual-use research in this section, during 2004 I certainly would have regarded this statement as a potentially fair representation of activities to that date.

Social analysis as constituting the dual-use dilemma

To engage in this or other lines of evaluative analysis of past initiatives requires taking a stance or making assumptions about a host of factual matters. Some sense has to be given of the real causes and consequences engendered by life science research in order to suggest what needs doing. While common enough in many policy analyses, many social scientists working within the 'social problems' tradition have questioned the appropriateness of what might be called an objectivist approach. Given that what is understood about the nature of any problem is the contingent result of the activities of individuals and groups (rather than a self-evident appreciation), a danger in taking (perhaps implicit) stances on the nature of problems is that social analysts end up making assumptions that should be questioned.

In what are generally referred to as 'constructivist' approaches, how claims about what the world is really like are made to appear 'objective' or definitive are taken as topics for analysis.[80] The notional focus is with the sometimes fraught processes of 'claim-making' rather than 'making claims' about the actual causes and consequences of social conditions. So rather than asking 'What are the "facts" of a problem?', attention is given to the ways in which claims are established as (more or less) agreed facts, used by certain individuals to identify a situation as a problem, and mobilized to suggest certain responses.

The choice between 'objectivism' and 'constructivism' has been a longstanding theme in the social problems literature. At stake is how and what to question. The choice, though, is not as simple as opting for one approach to the exclusion of the other. Rather, the two are intertwined.

Committing oneself to an objectivist examination when (as so often is the case) there is disagreement about the nature of the problem (e.g., with drug use, unemployment), the act of 'making claims' quickly gives way to attention to the processes of 'claims making'. Skeptical scrutiny is often cast, for example, on why certain claims are deemed credible, how definitions present a particular understanding of some phenomena, the historical reasoning for particular social priorities, the alternative ways of making sense of some given data, and so on. In short, the contingencies and constructed nature of any understanding become topics for discussion.[81]

Committing oneself to a constructivist examination of the formation of definitions and claims raises its own issues about how objectivist claims can enter into any examination. One is what to include in an analysis. Both the Introduction and Chapter 1 alluded to the contestation surrounding dual-use claims and the manner in which some accounts suggested contrasting points as pertinent. In fashioning their own portrayals, analysts – like those they study – must decide which accounts of events to draw on and exactly what is relevant about each. Those pursuing *objectivist* inspired analyses take as their central preoccupation the establishment of what claims are really valid; a difficult enough task in its own terms. For *constructivists*, though, the problem of selection and inclusion is even trickier. The open-endedness of what might be included raises concerns about how any description suggests a contingent understanding of an issue.[82] The previous accounts given of dual-use debates could have been substantially extended and incorporated a far more diverse set of issues. It also could have portrayed different issues at stake through subtle or not so subtle changes. From a constructivist stance, a danger is that what is included and left out owes much to the preoccupations of analysts. Concerns exist not just about what claims are included, but how analysts should orientate to claims that might have been raised but were not part of discussions.

Woolgar and Pawluch went even further by contending that while constructivist forms of analysis are notionally committed to moving away from talk of objective conditions, they often deploy the same type of selective questioning that is the subject of much constructivist criticism of overtly objectivist analyses.[83] This selectivity is built into the logic of constructivism, which assumes that the role of the analysts is to explain the indefinite relation between changing characterizations of a problem and the underlying condition. As such, assumptions about the nature of the underlying condition are often left uninterrogated. In this vein, for instance, the quote at the start of this chapter, that social problems 'may emerge, transmute, descend, disappear, or reappear, independent of any change in actual conditions',[84] requires being able to specify 'actual' conditions. Yet, it is not clear how the 'actual' conditions of some identified problem could be simply known. From a constructivist perspective, any understanding of these would itself be formed through the same contingent processes that inform the changing characterizations of the problem.

The difficulties in responding to the way in which objectivist assumptions enter into constructivist analyses can be illustrated by reflecting on the preceding argument of this chapter. With the exception of the previous section, I have largely avoided offering explicit claims about the nature of the dual-use potential of life science research. Rather, the chapter has cited and compared accounts offered by others. However, such an approach has traded on a taken for granted sense of there being some 'dual-use potential' that is being alternatively talked about. Yet following through the constructivist desire not to take characterizations at face value would caution against any such a practice. Moreover, while certain claims and definitions have been subject to scrutiny in the text above, the presumptions in the evidence used to support this questioning generally have been subject to much less.

These points raise the basic issue of what should be taken for granted and objectified in analyses that are committed to questioning how, and which, understandings become established. Some constructivists have responded to this predicament by calling for the avoidance of any reference to social conditions at all in favor of examining the rhetorical devices and styles of argument used to give warrant to claims.[85] In other words, the focus is on the nature of discourse rather than on whether statements refer to conditions in the world. However, such an approach itself relies on the potential of making a demarcation between the statements of analysts and 'non-analysts'. If analysts were regarded in the same way as others, then any of their claims could likewise be subjected to scrutiny regarding the contingent rhetorical strategies employed to construct their understandings of the discursive process of claim-making.[86]

Some analysts have responded to concerns about a creeping objectivism by acknowledging the limits of possible questioning in constructivist accounts, but then arguing that this is necessary in order to say something about social issues. To ask, for instance, why certain definitions of problems dominate policy debate requires moving beyond a narrow concern with rhetorical styles to talk about the structure of society at large.[87] Much here turns on the role and responsibilities accorded to researchers. Àlvarez made the case that social scientists:

> have a right, and perhaps even an obligation, to be participants in the process of identifying what constitutes a social problem, proposing remedies, as well as in evaluating societal responses and results. But we also recognize that our contributions to the creation of knowledge about social problems, as well as its use to induce social change, are, themselves, a moral enterprise. Our participation will, inevitably, either lend weight to, or detract from, the moral claims made by various contending constituencies within and between social systems.[88]

Since much will be taken for granted in any analysis – be that analysis trying to establish objective conditions, to unpack the contingencies of definitions, or to examine the styles of arguments – analysts should carefully consider what sorts of questions they ask and the justifications for assuming certain things.[89]

Concluding remarks

Whatever position one takes on the novelty or severity of dangers from life science research, questions are being asked in science policy circles about the direction and control of research with an intensity that was not there in the past. The then president of the ASM, Ron Atlas, captured many of the emerging issues being discussed in 2002 when he asked:

> Should scientists be constrained regarding questions they ask and should more research be classified? Should journals reject papers containing potentially sensitive information? Should secrecy clearances be required for attendees at biodefense research meetings? Should there be mandatory government review before publishing information, even from unclassified studies and those not funded by government? Finally, perhaps the most difficult questions of all, exactly what is sensitive information, and who is empowered to decide what is potentially dangerous?[90]

Answers to such questions raise basic issues about the place of science in society. In light of the significance of the issues being debated, major questions can also be asked about the conduct and purpose of research into dual use.

Chapter 1 has provided an initial suggestion of the choices and challenges in the examination of dual-use life science research. Those choices and challenges refer to substantive concerns about life science research as well as more conceptual concerns with the analysis of 'the problem' of dual use. For both, crucial questions can be raised about what needs doing and what is at stake in how research is approached. In the following chapters the distinction and inter-relation between what is malign and benign, taken for granted and questioned, novel and old, objectivist and constructivist, precautionary and unnecessary, and descriptive and evaluative will be revisited and incorporated into a program of research.

2
Discussing Science

As suggested in the Introduction and Chapter 1, concern about what science is like as an activity have been central to recent discussions about the relation between research and biothreats as well as the origins and scope of dual-use concerns. This chapter further considers the nature of 'science' to develop a fuller understanding of the security implications of research and possible responses.

In addressing these issues it continues the approach adopted so far of considering security issues while attending to what is at stake in the way analysis into this topic is undertaken. It does so by extending many of the initial points raised in Chapter 1 regarding the choices and challenges in social analysis. In that chapter, questions were raised about the distinctiveness and inter-relation of so-called 'constructivist' and 'objectivist' studies of social problems. As argued, while many analysts have recently sought to avoid using objective claims to knowledge to instead examine the origins and contingencies of definitions and characterizations, just how this refraining could be done remains a debated issue. This chapter moves on from a focus on making claims about social problems *per se* to more generic questions about the choices in what is taken for granted and what is questioned in analysis. Contrasting approaches adopted for the study of science are presented to map a constellation of possibilities in social research. Attention to what is taken as a topic of analysis and what is taken as a resource will be further addressed in Chapter 3, where a rationale for a particular investigation into dual-use issues is outlined.

Prevalent models of science

Various assumptions about the nature of scientific practice have factored in post-9/11 security discussions. The presumption that science is characterized by openness has been pervasive, as too have been claims about the importance of freedom. Such 'social'[1] dimensions to science have been said to limit the appropriateness of security controls since any restrictions would undermine the advancement of knowledge. Openness has been said to be essential for gathering facts, validating conclusions, and scrutinizing claims. Questions voiced about what to do about sensitive findings have regularly been framed in terms of the difficulties of 'balancing security and openness in the conduct of research'.[2] Reiterating previously noted statements, one prominent defense analyst commented that:

> The frightening possibilities [of the novel bioweapons] taken together with the difficulty of regulating or controlling such potential adverse outcomes outside the laboratory, suggest the need to consider mechanisms for the governance of fundamental biological and biomedical research. However, governance mechanisms based on the potential application of a given line of research are difficult to reconcile with the established practices of the international scientific community, which attaches great importance to academic freedom, to free and open inquiry, and to protecting an individual investigator's ability to let scientific promise alone determine the direction in which research should evolve. These community norms have paid off handsomely, as evidenced by the explosive progress in science and technology – progress that has been fueled by numerous scientific breakthroughs that could not have been predicted, planned, or managed.[3]

A report by the US Congressional Research Service concurred with some of these sentiments in stating: 'scientific enterprise is based on open and full exchange of information and thrives on the ability of scientists to collaborate and communicate their results.'[4] The NRC report on access to genetics research, *Seeking Security*, similarly contended that 'Unfettered, free access to the results of life-science research is the historic norm and has served science and society remarkably well. Open access allows life scientists everywhere to evaluate, interpret, adapt and extend results from many fields of inquiry for use in their own work and thereby accelerates research and speeds the delivery of life-saving benefits that biological and medical research are so rapidly creating.'[5]

Statements made as part of dual-use discussions about the role of openness, freedom, and norms do not propose to set out comprehensive accounts of the practice of science. However, they share language and conclusions similar to certain well-known systematic analyses.

Robert K. Merton's classic essay 'The Normative Structure of Science' (1942) sought to identify the characteristics of science that could account for its functioning and rapid progress in modern times.[6] Principal among them were social norms governing the conduct of individual scientists through community incentives and sanctions. These included expectations such as impartially evaluating claims on the basis of their merits rather than extraneous factors, pursuing research for the advancement of knowledge rather than personal gain, as well as the open sharing and common ownership of research.

Assumptions made about the nature of science also factored in post-9/11 discussions in more indirect ways. As noted, questions about what to do about dual-use research findings and techniques have often been framed in risk–benefit terms. Herein the call has been made for an expert-led process of identifying the risks and the benefits of individual publications or research projects. These competing factors are to be weighed against one another to derive an overall assessment for action.[7] The manner in which it was suggested that opposing factors be identified and evaluated mirrors many conventional assumptions about how evidence functions in making scientific conclusions about risk. Central to such rationalistic procedures is the ability of scientific experts to determine the significant considerations and offer authoritative appraisals. While there has been little suggestion that all scientists necessarily will come to the same appraisal about the dual-use potential of research, expert conclusions are not assumed simply to derive from the vagaries of individuals. It should be underscored that it has been *technical* expertise that has been widely presented as essential to offering assessments for action. To the extent mention has been given to the public or political representatives in dual-use policy discussions post-2001, these have tended to be warnings about the dangers of their involvement stemming from a lack of technical understanding.

Science under the microscope: norms and facts

In the last few decades, varied lines of work undertaken through social studies of science has sought to offer an understanding of science as a social activity. This section surveys a number of such analyses with a view to considering their implications for the assumptions about science noted previously in dual-use policy discussions. Differences in the studies mentioned will also be used to consider what is at stake in different types of social research.

Norms under threat?

If suggestions about the nature of scientific practice as being open or free are taken as accurate, then any attempts to constrain research would have serious consequences. Are such notions correct though? In science policy analyses, much attention has been given to limits of openness and freedom in recent times because of the said increasing commercial orientation of public research.[8] Particularly with regard to the life and medical sciences, concerns have been raised that scientists are no longer primarily motivated by peer esteem and recognition, but rather financial reward looms large. For instance, professional organizations partaking in biosecurity discussions such as the US National Academies or the British Royal Society have themselves warned about the corrosive effects of commercial calculations on the degree of openness and sharing among scientists.[9]

Krimsky and other analysts[10] have pitched their concerns about the impacts of commercialization in terms of a threat to the norms of sciences. For him, the ever-increasing pursuit in the United States of knowledge because of its monetary value has lead to 'significant changes in the culture, norms and values of academic science'.[11] These changes have resulted in an undermining of the traditional Mertonian norms and ethos of science. By drawing on a wide range of prominent cases involving conflicts of interest, questions of bias, and decisions about the publication polices of journals, Krimsky argued that by 2003, US universities had regrettably become a 'different type of institution'.[12]

While concurring with such broad conclusions about the constricting effects of greater propriety controls for the exchange of scientific information and materials, other analysts have done so without seeking to offer an overall evaluation. MacKenzie, Keating, and Cambrosio, for instance, assessed the introduction of patents in monoclonal antibody research and contended that they posed severe restrictions on both the flow of information and researchers' activities.[13] Yet, they did not lament this development, but rather raised it as an issue for scientific professionals to 'reflect upon'.[14] Of course, in the overall backing of commercialization practices, universities and other organizations have often contended that any possible negative implications of particular commercial ties on the norms of openness and sharing are more than compensated for by their benefits.[15]

Most of the analyses cited in the paragraphs above make direct reference to a transformation in the norms of (academic) science. Others within science studies have long sought to question the extent to which science is characterized by freedom, openness, or any other such norms.

At least since the 1960s, substantial critiques have been offered to the suggestion that science ever functioned in accordance with dominant norms such as openness, disinterestedness, or communality – this including critiques made by Merton himself.[16] In a study of Apollo lunar mission scientists, Mitroff called conventional accounts of science as norm-guided part of a misleading 'storybook image of science'.[17] For every supposed norms of science (such as disinterestedness and impartially evaluating claims), he argued that one could find scientists citing the importance of so-called 'counter norms' (such as emotional involvement and the evaluation of claims because of their source). While openness is often said to be central to science, secrecy was seen by those interviewed as minimizing disputes about who has discovered what while ensuring standards of reliability. The alternation between conflicting norms was not a cause for criticism for Mitroff, but rather a necessary social condition for individuals trying to cope with contingent and complex situations.

In 1975 Mulkay went further to argue that suggestions that science operates according to any norms (or counter-norms) were fundamentally inaccurate because there was little indication that the norms often cited were incorporated into reward or sanction structures. Rather than norms providing the central basis for determining conduct, they functioned as a selective 'vocabulary of justification' for projecting an image of science to outsiders. This image was 'positively misleading, but it served many scientists' professional interests.[18] Mulkay justified this conclusion about the ideological role of norms in large part by drawing on historical studies of the emergence of the notion of 'pure science' in the United States during the nineteenth and twentieth centuries. This notion being that science should be publicly supported without the normal public controls on where and how funds were spent. The consolidation of this way of thinking about science in government was said to owe much to a selective rhetoric of the 'norms of science'.[19]

In a parallel manner to the analyses above regarding the commercial and competitive restrictions on science, the potential introduction of security-inspired restrictions could be approached in a number of ways: a further deterioration of the traditional norms of science, a subject for detached assessment, a welcomed introduction of new considerations into the research process,[20] or an occasion for playing out contests of professional autonomy.

Analyzing arguments about and in science

Many of the analyses of norms cited in the last section were based on primary or secondary interviews with scientists. Through identifying commonalities across statements and generalizing from these, analysts justified conclusions about the changing norms of openness, the role of non-cognitive factors in research, the political implications of certain notions of science, and so on. In a significant departure from his analysis of norms-as-ideology, in the mid-1980s Mulkay, with Gilbert, offered a warning about the methodological appropriateness of many types of social research of science.[21]

The basic starting point for this was the potential variability of scientists' accounts. Based on extended interviews with biologists in the field of bioenergetics, Gilbert and Mulkay argued that scientists often spoke about their practice in multiple and divergent ways depending on the context in question. Deriving definitive analysis from such accounts requires sorting through an often considerable amount of conflicting statements to determine what really holds.

They illustrated this variability through contrasting different 'interpretative repertoires' used by scientists. In writing for research articles, the biologists studied overwhelmingly employed a so-called 'empiricist repertoire'. This meant providing highly impersonal accounts devoid of any concerns about the preconceptions or commitments of the authors. Facts given were not presented as reliant on interpretation, but rather as the natural outcomes of closely studying the world. Yet in more informal settings, such as interviews, this conventional way of talking about science was *mixed* with a contingent repertoire wherein 'scientists presented their actions and beliefs as heavily dependent on speculative insights, prior intellectual commitments, personal characteristics, indescribable skills, social ties and group membership.'[22] Under this contingent repertoire, scientists' statements were not presented as straightforward, self-evident representations of the world.

Gilbert and Mulkay took the variability in accounts as indicating the dangers of social analysts trying to offer some definite truth about what really takes place in science from the selective use of interview or observational data. To do so would require special knowledge about which repertoire for describing science was really accurate. Rather than pursue this task, they argued that analysts should attend to how different ways of accounting for science are rhetorically organized in certain settings (e.g., articles, interviews, speeches) to give warrant to claims. This suggestion parallels that in Chapter 1 advocating a 'strictly' constructivist view of social problems in which analytical focus turns away from the actual nature of problems to the nature of discourse.

For instance, Gilbert and Mulkay asked how scientists are able to account for disagreement about scientific facts given the empiricist repertoire so often employed. The problem was that 'Each speaker who formulates his own position in empiricist terms, when accounting for [others'] error sets up the following interpretative problem "If the natural world speaks so clearly through the respondent in question, how is it that some other scientists come to represent the world inaccurately?" ' Those interviewed shifted between describing scientific research in empiricist and contingent repertoires, depending on whose claims were being assessed. Non-experimental, non-scientific factors were regularly utilized to account for the 'errors' of others who maintained competing theories and conclusions.

This basic orientation was later extended and transformed by Mulkay in a study of replication in science.[23] In the area of bioenergetics he examined, whether certain experimental results were repeated by others was a complex matter. Rather than the replication of results being a straightforward way to test the validity of claims, the sameness or difference between certain findings depended on various interpretations. Scientists disputed whether they were, in fact, in dispute about the data or the conclusions of experiments. Experimental work, for instance, might be presented as indicating findings that were different in detail but essentially the same in substance. Since the scientists in question rarely tried to conduct near exact duplications of others' work, many claims about replication rested on the ability of some researchers to use others' results to make predictions relevant to their different work. In short, 'replication' was a thoroughly social accomplishment.

And yet, in offering these conclusions, Mulkay recognized that his argument itself mobilized claims about the similarity and difference between scientists' statements to discuss how they used notions of similarity and difference to talk about replication. Rather than just scrutinizing the way scientists distinguished facts from untruths, he also scrutinized how he did the same in his analysis. This was done, in part, through blending 'factual' and 'fictional' styles of argument that sought to highlight and question his use of facts, determinations of similarity and difference, and impersonal 'empiricist' forms of writing.

The varied points given in this sub-section about the practice of science were not offered by Mulkay or Gilbert as a sign of some deficiency in the conduct of scientists or the unreliability of scientific knowledge. Instead, much of it was geared to critiquing the methods employed by those studying science. In yet another turn though, in his later work Mulkay asked whether communication between scientists could be improved. As he argued, informal communications between scientists about areas of disagreement can be unproductive in resolving differences when those exchanges are characterized by the empiricist repertoire that appears in formal texts. This is because, in such exchanges, individuals presume the factual and obvious status of their own claims, while dismissing others' beliefs as fiction resulting from non-scientific factors. This made it difficult to establish a dialogue in which the interpretations and reasoning for appraisals could be jointly considered. Offering an evaluative analysis, though, required Mulkay to go beyond the detached study of the rhetorical organization of discourse. Instead, assumptions about communication had to be employed and standards set for what constituted good communication.

Analysis as critique

This chapter began by highlighting many of the themes of the Introduction and Chapter 1 regarding the said importance of norms of openness and expert-led appraisals. A brief survey of the varying ways analysts have argued about the norms of science led into a methodological discussion that raised issues about how scientists and social analysts use evidence. With regard to both topics, key questions can be asked about the purpose of analysis as well as what is taken a as topic for study and what is taken for granted. Mulkay's diverse studies of science illustrate a variety of possibilities for handling statements about the nature of science: they can be resources for making arguments about the world (e.g., talk of norms serves as ideological cover against attempts to make science more accountable); they can be taken as topics of study in their own right while avoiding concerns about their real-world validity; or some 'middle ground' can be pursued of treating statements as both 'about something' but also as patterned forms of talk that owe much to the situation in which they are given.

This 'middle ground' contains a variety of possibilities. For instance, Kerr and Cunningham-Burley drew on the work of Gilbert and Mulkay to attend to the way in which geneticists in the United Kingdom employed interpretative repertoires and discursive strategies to discuss the relation between 'science' and 'society'.[24] Because scientific knowledge is often said, on the one hand, to be neutral and objective and, on the other hand, to have profound social and ethical implications, how scientists relate 'science' and 'society' can often be problematic.

Kerr and Cunningham-Burley sought to understand how geneticists made this connection by attending to wider concerns about social power and professional interests. A central aim was to critique the opportunistic ways that boundaries were drawn between science and society.[25] As they contended, geneticists often variably and flexibly defined the science–society relation, to maintain their authority and status but also to distance themselves from the responsibility for any negative implications of genetics. So, while it was recognized that science provided knowledge that could be misused (as in genetic discrimination), any such misuse was the fault of 'society' not 'science'. However, because geneticists were often involved in advising governments and determining policy (e.g., with regard to genetic testing for insurance purposes), acknowledgement was sometimes made that geneticists had some responsibility for ensuring the appropriate use of genetics. That responsibility, though, was seen as limited to supplying neutral and impartial information. Yet, as Kerr and Cunningham-Burley argued, claims about the neutrality of advice could only be maintained by geneticists ignoring that they had definite professional interests in the policy advice given and decisions taken. Through such discursive work, many geneticists sought to occupy a vital role in educating society and defining policy concerns, but without accepting responsibility for the negative social outcomes of genetics.

The validity of analyses such as those by Kerr and Cunningham-Burley rests on the acceptability of employing a given sense of professional interests as a resource to interpret and critique geneticists' statements. This assumption is taken as a basis for making sense of contrasting interview data and a platform for evaluating the appropriateness of responses.

Analysis as understanding

Yet another way for thinking about what analysts should take as a topic of study and what they should take as a resource is provided by Collins's examination of attempts to detect gravitational waves.[26] His study of the gravitational wave community, spanning some 30 years, sought to provide a sociological analysis of how claims and theories were justified, how this field developed from its initial origins, and how during that development the many uncertainties and disputes about scientific facts were addressed. For instance, Collins detailed differences in standards for interpreting experimental evidence between two research teams concerned with the detection of gravitational radiation. While one sought to publish somewhat tentative findings quickly in order to further wider scientific debate, the other sought to ensure the highest level of confidence in any conclusions reached before publication. He related these alternative approaches to national funding patterns and histories with gravitational wave research.[27]

In the pursuit of such matters, Collins did not make evaluations of whether certain scientific claims were true or false. Rather the approach was one of 'methodological relativism'. However, in considering how claims became to be regarded as facts, only certain facts could be scrutinized. In a manner that might be treated as paralleling certain constructivist approaches to the study of social problems (Chapter 1), rather than seeking to question the basis of all factual claims he had to treat certain matters as given in order to selectively examine the basis for others.

The questions of what should be taken as given (in Collin's words as part of the 'scenery') and what should be a topic for analysis (the 'action') were directly linked to the audiences for research and the competences required by analysts. To speak in a way that could bring novel sociological insights to bear on the work of scientists required assuming some (but not too much) distance from their ways of thinking. So, he said, analysts have to question matters (such as the criteria for judging the validity of findings) which scientists often took for granted. Though criticism was not an aim of the type of analysis Collins pursued, it could often seem that way for scientists. If the extent of analysts' questioning becomes too broad though, it risks raising issues that are no longer relevant to scientists' work.[28]

Despite acknowledging the importance of what is taken as scenery or action, Collins provided relatively little elaboration of why certain choices were made in his research instead of numerous others. Whereas authors such as Kerr and Cunningham-Burley justify their choices as stemming from a concern to raise critical questions about the social implications of genetics, he offered little by the way of explicit strategic rationale. So, Collins contended, 'where to treat scientific findings as a matter of fact and where to treat them as a matter of analysis is my choice – a matter of my judgment.'[29]

Social studies of science under the microscope: topics and resources

The previous section provided a brief overview of certain social studies of science pertinent to the themes offered in discussions about the dual-use potential of research. These pointed to, first, the potential for diversity in scientific practice and, second, the diversity in how analysts examine science.

With regard to the first, rather than talking about the practices of 'science' in general, it is necessary to talk about the highly varied practices of science across sub-disciplines and institutions.[30] As well, it was suggested that the marshaling of evidence to make convincing scientific arguments (regarding the replication or the validity of facts) should be approached as negotiated accomplishments.

With regard to the diversity in analysts' approaches, the studies considered above provided contrasting purposes for the social studies of science. These included understanding the dynamics of the development of a field, critiquing how scientists attempted to resist forms of accountability, engaging with researchers to improve their communication, and studying how language works. Each entails a different way of making arguments, each is likely to appeal to certain readers, and each draws on particular types of data in support of the conclusions offered.

As has been argued, a vital matter faced in studying science is what is taken for granted (part of the 'scenery', a resource) and what is questioned (part of the 'action', a topic of analysis). While hardly exhaustive of the types of approaches adopted for the social studies of science, the last section did chart a wide spectrum of responses to this basic matter. Following what was referred to as an 'objectivist' agenda in Chapter 1, some analysts have sought to make definitive claims about the state of the world by drawing on certain facts. Krimsky, for instance, took the historical centrality and value of norms as given in order to claim that the situation today in the medical and life sciences (regrettably) had deteriorated. The early work of Mulkay sought to question whether norms did structure the practice of science in the way so often claimed and in doing so presented nearly an opposite evaluation of norms to Krimsky. Both, though, supported their argument through attending to fairly large-scale historical and institutional changes.

Instead of explicitly making claims about what is really going on in the world through drawing on accounts of what science is like, other analyses have instead been preoccupied with claim-making processes. Gilbert and Mulkay's attention to discursive repertories necessitated a detailed examination of copious arguments from interviews and publications. In doing so, they raised significant doubts about the appropriateness of simply drawing on certain accounts of scientists as straightforward resources to make claims about what was going on in their world. Yet, the divide between those analyses that seek to 'make claims' and those that study 'claim-making' is by no means clear-cut. While Gilbert and Mulkay were concerned with how biologists' discourse functioned in claim-making, their analysis of these practices treated their own arguments about science in a fairly unproblematic manner. Later Mulkay would subject both the arguments of scientists and himself to scrutiny in order to undermine taking certain claims for granted. As he acknowledged, however, such a reflexive questioning of claims could never be total.

In order to say something about the social consequences of modern genetics, Kerr and Cunningham-Burley took the professional interests of geneticists as an analytical resource to make sense of the way in which geneticists constructed the relationship between science and society. Drawing on concept of interests derived from macro study of social institutions in this manner provided the basis for critique. In order to say something about the development of a scientific field, Collins took certain scientific claims as given in order to question the evidential basis for others.

Reconsidering dual-use framings

In Chapter 1, the broad contours of dual-use research as a social problem were considered and through these, questions were raised about whether analysts should seek to offer authoritative definitions about the nature of social problems or instead attend to how such problems come to be defined. This chapter has extended these themes by considering the underlying assumptions about the nature of scientific practice in dual-use debates and through this wide-ranging questions have been posed about what gets questioned or taken for granted in analysis. The remainder of Chapter 2 reconsiders what counts as troublesome dual-use research in the light of the previous argument of this chapter.

Just the facts, please

As mentioned in each of the chapters so far, many discussions of proposed controls for dual-use research rely on the possibility of distinguishing 'problematic' from 'unproblematic' knowledge and techniques; this mainly through weighing their benefits and risks. Such discussions typically stress the importance of engaging in expert-informed processes of identifying the risks and the benefits of publications or projects. In a manner consistent with 'objectivist'-type approaches to social problems, those adopting this way of thinking about dual-use issues seek to set out definitive determinations of what is going on. Such arguments mobilize facts as resources for justifying particular evaluations.

Despite the importance often attributed to establishing fact-based procedures to determine where the correct balance rests between openness and security, as of mid-2006 little has been offered by the way of detailed (let only quantified) analysis of the relative risks and benefits of research. Concerns and promises associated with certain bits of research have been identified and the conclusions of the weighing announced, but generally with only limited elaboration of the actual process for making evaluations. The most substantial discussions have resulted from relatively high-profile, contentious decisions that attracted commentary from various quarters.

As discussed already in Chapter 1, the sequencing and reconstruction of the 1918 Spanish Flu was one such high-profile case. Although few commentators cited as part of reports in the scientific and mainstream press came out against the undertaking and publishing of this work, these actions were subject to much debate. One of the most (and one of the very few) high-profile categorical statements against the publishing of the DNA sequence data was given by Kurzweil and Joy in an article for the *New York Times*.[31] They likened the genome to a 'recipe' and a 'design' for a weapon of mass destruction and thus suggested its publication was irresponsible. In fact, these authors contended, the sequence data was even more dangerous than the design of the atomic bomb because of the contagiousness of Spanish Flu and the relative ease of synthesizing it. Instead of openly publishing the sequence information, Kurzweil and Joy advocated restricting key information about the sources of virulence or establishing security clearances for access to some information.

Underlying their evaluation of the 'foolishness' of the sequence publication was an assessment of the relative ease by which such scientific findings could be turned into a dangerous weapon. Whether this is accurate has been a matter of some discussion. One *New York Times* letter offered in response reversed the ease Kurzweil and Joy attributed to synthesizing complex virus and (in this case) enriching uranium.[32] Another cited the potential for misuse of only certain information about Spanish Flu to argue against limited restrictions since even partial sequence data could be misused.[33] Still another letter stated that the data serves as a 'formula' for a weapon, but did so to then argue it was foolish for Kurzweil and Joy to further publicize the potential implications of the research through writing widely accessible articles about it![34] In relation to arguments mentioned in the Introduction, whereas Kurzweil and Joy took for granted the contagiousness and deadliness of any recreated virus, others doubted such conclusions by querying whether the single genome pieced together was actually a highly deadly variant (see page 18) or noting that the evolution of human immunity meant any synthesized virus was unlikely to be as deadly today as it was a century ago (see pages 8–9).

Thus, points of disagreement have been evident in what limited consideration has taken place about the benefits and risks of research in public discussions. To the extent it is voiced, such disagreement problematizes the potential for the straightforward determination of what should be done. In such a troubled state, further questions are likely to be asked: What criteria are used for making decisions about the risks and benefits of research? What counts as credible evidence and who is a credible commentator? How should those making decisions handle any uncertainties? What significant matters for consideration are uncertain?

Arturo Casadevall, a member of NSABB, spoke to many of these concerns at a presentation to the first meeting of the Board in 2005. This was given as an initial contribution to its deliberations regarding the criteria for defining what counts as dual-use research.[35] As he contended, determining the effectiveness of any bioweapon was 'dependent on both the microbe and the host, and many of the interrelationships are not understood'. That host-agent interdependency undermined developing clear-cut classifications of what might be dangerous because it hindered making easy generalizations. This meant that (in theory) almost everything might be deemed dual use. Even everyday products, such as those in yogurt, could be dangerous for certain individuals. Casadevall argued that efforts to move beyond noting the ever-present danger of bioagents to specify what was really of concern were frustrated because we 'lack the basic information to make weapon potential calculations even with this very simple type of relationship for most of the agents that are already known to be pathogenic' (i.e., the relationship between the virulence, communicability, stability, and the time till impact of a microbe). He concluded with self-identified 'personal' remarks that where the line was drawn about what to regard as of concerns and thus potentially in need of control was ultimately 'political', this in the positive sense that any red lines were matters for deliberation.

Disagreements about matters of fact evident in prominent discussions about dual-use issues suggest the attempt to achieve definitive resolutions of the relative dangers of research is likely to have some limitations. Unless one simply assumes certain facts are beyond doubt (if only certain misguided individuals would think with sufficient rigor), what counts as 'troublesome' research and why should be thought of as a topic of possible contestation. How much consensus about the assessment of such research could be generated through enhancing existing knowledge about bioagents' characteristics, host susceptibility, or the sources of attacks is an important and interesting question, but one itself unlikely to be a matter of unanimity.

Making claims and claim-making in the analysis of dual-use research

How then might the types of social analyses of science discussed in the previous section of this chapter inform the matter of what counts as of concern, contentious, and so on dual-use research? Against the potential for dispute noted in the last section, one approach highly indebted to an 'objectivist' agenda would be for social analysts to offer definitive assessments about the dual-use status of research. This could be accomplished by drawing on certain types of scientific arguments as resources to contend what (by some measure) is or is not really problematic. On just what basis social analysts could seek to credibly resolve what are typically defined as highly scientific disputes would no doubt be a key issue. Alternatively, they might appeal to a different form of expertise to justify the authority of their appraisals. Knowledge of the history of biowarfare and bioterrorism, the relation between scientific discovery and practical innovations, or an understanding of the governance of science might be such bases. A related aim might not be to make assessments of individual elements of research, but to comment on the framing of the overall debate.

The attempt to resolve disputes, though, is just one possible contribution that might be offered. As outlined earlier, many analysts of science have taken as their concern the manner in which alternative claims are made about the utility or novelty of research. Herein, disagreement about research is not treated as somehow atypical or unfortunate, but a phenomenon for examination itself. Casadevall's evocative conclusion regarding the importance of the 'political' in determinations of appropriate research invites attention to claims-making. As argued previously though, locating the emphasis of analysis with 'claims-making' rather than 'making claims' is only a preliminary move in deciding what analysis to offer. Many further complicated issues about what is questioned and what is taken for granted must be addressed.

The importance of attending to claims making and the further choices it demands can be illustrated through examining claims about the IL-4 mousepox experiment. At a meeting organized by the US National Academy of Sciences, the then editor of the *Journal of Virology* and President-elect of the ASM, Thomas Shenk, spoke on why the journal had published the article. Much of the justification was framed in terms of the difficulty of identifying what constituted problematic research because of the incremental nature of science.[36] As elaborated, to restrict research about the effects of IL-4 in mousepox would be ineffective since it was a small step from research on other pox viruses to this particular one. To restrict research about the effects of IL-4 into other viruses than mousepox would undermine knowledge of how IL-4 functioned. To restrict research about the discovery and general functioning of the IL-4 gene would impede work into fundamental mechanisms in immunology. Yet, even such basic information might be enough to get some guessing about the possible malign implications of IL-4. In the end, he concluded that attempts to draw restrictions around research that might suggest how IL-4 could be misused would 'seriously risk disrupting the foundation on which biomedical science progress is building'. Through this sort of argumentation about the incremental nature of science advancement, Shenk argued biosecurity attention should be given to those lines of research that resulted in unexpected findings (of which the IL-4 mousepox case was *not* an example).

Analysis as understanding

An analysis of 'claims-making' about the status of research might take a number of forms. A basic starting point in this would be noting that assessments of the (benign or malign) utility of research are not determined simply by the experimental facts of particular lines of research. Herein, just where a 'red line' should be drawn for research would be approached as depending on the contingent criteria for evaluation, what counts as relevant data, and the context for consideration. In short, establishing lines would require much interpretative work. Following research such as that noted above conducted by Collins and others, one might turn attention to the (often implicit) standards and procedures employed for making sense of the research results.

Because of the limited articulation of decision-making about dual-use research, investigating those standards on the basis of existing publicly available information would be difficult. One possible future method for investigation to get at these standards, though, would be to compare the manner in which claims are made about the contribution of research to malign applications along with the contribution to benign ends. To elaborate, Shenk contends that restricting research findings involving IL-4 relevant to bioweapon applications would 'seriously risk disrupting' the basis of biomedical science since doing so would catch great swathes of activities. This was because results such as those obtained by the Australian researchers could already be 'guessed at' from what, as he said, was 'out there in the literature'. Yet, particularly because of the relative prestige of the *Journal of Virology*, this raises the question of the reasoning for publishing the article at all. Unearthing the basis for such contrasting appraisals might well bring to the fore many taken for granted assumptions. Examining contrasting ways of making sense of 'the same' experimental results might also help make sense of the puzzle identified in Chapter 1: while it is widely recognized that all knowledge and techniques have a dual-use potential, examples of problematic dual-use life science research in recent biosecurity discussions have been quite limited.

Following on from the points raised in this sub-section and those made earlier in this chapter regarding the negotiated status of the 'replication' of results, it seems unlikely Shenk's proposal to direct biosecurity attention to 'unexpected findings' (or findings 'immediately' relevant to bioweapons[37]) is likely to be a straightforward one. Shenk himself acknowledged disagreement about the predictability of particular aspects of the IL-4 mousepox article.[38] In general, the meaning of 'unexpected', 'immediate', 'new', 'unique', or any other characteristic would be open for dispute. For instance, on the back of research carried out in Australia on IL-4 and mousepox, a research team in the United States applied similar procedures to cowpox. A rationale given for this was 'To better understand how easy or difficult it would be to apply the same kind of genetic engineering to the human smallpox virus and make it more lethal.'[39] According to reports though, one of the Australian scientists questioned the need for this additional work given that they had found the insertion of IL-4 worked as expected for rabbitpox. As suggested in the news report, since cowpox can affect humans and the research would yield little by the ways of new findings, there were significant questions about its merits. It seems reasonable to assume that such contrasting appraisals about research are based on unspecified differences in evaluations such as what certain experiments really indicated, what was already known, and what could reasonably be inferred.

Another way to examine standards for interpreting scientific evidence would be to compare decision-making processes across national contexts, where broad differences might emerge. For instance, whereas in the United States efforts are currently being made through NSABB to derive detailed formal standards for making appraisals, no such official efforts are underway in the United Kingdom. When the director of the UK-based Wellcome Trust was asked in 2004 what criteria reviewers of grant applications have used to judge issues of dual use, he refused to give a 'formulaic answer because I think these are all matters of judgment .... I think it is difficult to define any precise formulaic answer to that. I think they use their wisdom.'[40] Arguably such contrasting styles of decision-making suggested in these alternative approaches are not just differences peculiar to this topic at a given movement in time, but instead are indicative of general contrasting national risk regulatory approaches. Whereas British public agencies generally adopt informal, fairly closed processes that rely on the qualities of decision-makers for credibility, in the United States much more demand is placed on devising explicit criteria that weigh notionally objective risks and benefits, open processes, and credibility generated by formal procedures.[41] Such differences in approaches might lead to significant differences in what decisions are made and how they are publicly justified.

Analysis as critique

Although the issues raised in the previous sub-section might be taken as an implicit criticism of how determinations are made about the status of research because they pointed to possible inconsistencies, this need not be an aim or outcome. It is always possible, however, to quickly turn inconsistencies or divergences into an accusation. Following the example of Kerr and Cunningham-Burley, for instance, one approach that might be taken is to explain the way assessments are made by appealing to the interests of those offering them. Herein, the pattern of classifications for research would be scrutinized for how it served certain goals. For example, in relation to Shenk's incrementalist justification for the IL-4 experiment, in a 2003 article I argued:

> The attempt is made here to have it both ways: experiments such as the mousepox one are simultaneously deemed significant intellectual advances which merit publication despite possible contention but also insignificant achievements due to previous work in the area. There are two dynamics at work here, one whereby facts get built up and another whereby attribution of responsibility and contribution for producing facts are allocated. The two acting together enable individual scientists to be portrayed simultaneously as both players and pawns regarding the consequences of research.[42]

As it was more or less explicitly argued, this contingent asymmetry in accounting for research done by Shenk was useful in distancing scientists from any negative consequences of their work, while maintaining traditional forms of scientific recognition and also ensuring the debate remained largely a highly technical (and therefore highly exclusionary) one.

Analyzing arguments about and in science

Both the attempts to 'understand' and to 'critique' rely on mobilizing *certain* accounts made by scientists and others as argumentative resources for stating what is really taking place. In this, as contended previously in this chapter, a selective process takes place of questioning the basis for some dual-use type claims while taking others at face value. Yet, in following the suggestion by Gilbert and Mulkay, one might instead seek a more robust treatment of statements as topics for analysis. This could be done by shifting away from using them to make claims about what is really going on in the world to first (or only) examining the ways in which these statements are rhetorically organized, patterned, and given in specific interactional contexts with certain implications.

For instance, the reference I made above to Casadevall's appraisal of the central role of the 'political' in assessing research was used as a justification for attending to the contingency of claims-making about the status of research. Yet, in doing so, this statement making a point of indeterminacy was treated in a taken for granted, straightforward manner. The appropriateness of this might well be doubted. Rather than taking such a quote as indicating a state of high uncertainty, it could be taken as a single instance of (at least potentially) different ways of accounting for research. So, in line with Gilbert and Mulkay, this personal, and in many respects non-traditional, statement might be seen as an instance of a 'contingent repertoire'. What would be needed from there was a detailed investigation of how Casadevall and others make (and mix) such statements with those in line with a traditional 'empiricist repertoire' where the facts speak for themselves. A starting point would be contrasting the likely more formal language and arguments used in journal articles or other written analyses with that offered as part of oral presentations or newspaper articles. Whether or not 'repertoires' are taken as a proper driving analytical concept, through a comprehensive analysis of a wide range of scientists' and policy-makers' accounts it would be possible to comment on the basic dynamics of discourse.

As a footnote to what has been said in this section, any of the approaches outlined here that examined the claims making procedures in science could be turned back on themselves. So, while the social analysts discussed how scientists negotiate standards for validating research, make claims about the similarity and dissimilarity of research results, and question some things while taking for granted, these are all considerations that apply to social analysis as well. This chapter itself has offered claims about what 'following' or being 'in line' with the analytical approaches outlined might mean. The basis for such said relations could be elaborated and scrutinized. As might be argued, following similar points made above regarding the role of analysis in understanding, such scrutiny need not undermine any such claims. Whether and how social analysts incorporate such a reflexive 'turning back' task into their analysis speaks to how they formulate an understanding of their analytical responsibilities and what distinctions they make between the status of their accounts and the accounts of those they study.[43]

Closing remarks

This chapter began by examining some of the common presumptions about the nature of scientific research prevalent in dual-use discussions. A consideration of the norm of openness and the role of evidence in science overall led to a consideration of the different normative, methodological, and empirical ways social analysts have orientated to claims about, and in, science. Such contrasting approaches were then in turn re-employed to propose alternative ways of thinking about and investigating the potential of the dual use of research. As suggested, there are ever-present concerns in analysis about what is taken for granted and what is questioned. The next chapter moves on from these issues to outline a rationale for empirical inquiry.

3
Inquiry, Engagement, and Education

Earlier chapters considered the choices and challenges for social and life science research in relation to the destructive potential of modern bioscience and biomedicine. In these, the question of 'What should be done?' loomed large. Chapter 2 reviewed many of the concerns expressed regarding what effect policy responsive measures might have on the character of scientific research. The presumptions about how science operates underlying voiced concerns were then examined through considering a range of approaches in social studies of science. The inter-relation and diversity of such studies extended attention given in Chapter 1 to what gets taken for granted or questioned as part of analysis.

This chapter sketches a rationale for research about dual-use issues. It asks what could be sought from social research and what types of research might be promising in achieving such goals. The case is made in this chapter for a problem-orientated strategy that aims both to collect information about the reasoning of practicing scientists and to engage with them in a process geared towards mutual learning.

Moving forward with education?

While the previous chapters have repeatedly characterized the dual-use potential of research as a tension ridden topic, the policy discussions about it have been characterized by disagreement in every aspect. At least for those partaking in many high-level Western policy and scientific deliberations, there is fairly widespread accord on the need to pay greater attention to the destructive potential of advanced life science research. That attention has generally been seen as something that should be extended beyond traditional concerns about the storage of pathogens or the vetting of lab personnel. Instead, it should include the possible consequences of the data, conclusions, and techniques of research. However, underlining such general agreements are differences regarding the severity, source, and the likely targets of biothreats. So, too, there has been widespread agreement about the need not to jeopardize the said numerous benefits of science, but also varying conclusions about how this could be achieved. Some have contended that life scientists must act to pre-empt future draconian measures initiated from elsewhere; others that the introduction of controls threatens to be the thin edge of an ill-advised wedge.

The need for further education and awareness raising of scientists about dual-use issues is another topic where the wide ranging agreement at a general level arguably gives way to disagreement at the level of specifics.[1] General calls for increased education have figured into numerous recommendations about what needs doing since 9/11.[2] Extensive familiarity with dual-use issues within the life science community is vital if that community is expected to largely self-govern itself. In relation to concerns about biological weapons, in 2002 the British Royal Society said: 'Consideration should be given to some formal introduction of ethical issues into academic courses, perhaps at undergraduate and certainly at postgraduate level.'[3] A meeting of scientists, policy-makers, and others in 2004, convened by the Royal Society and the Wellcome Trust, concluded: 'Education and awareness-raising training are needed to ensure that scientists at all levels are aware of their legal and ethical responsibilities and consider the possible consequences of their research.'[4] Internationally, the importance of furthering awareness was highlighted in a 2005 Meeting of Government Experts to the Biological and Toxins Weapons Convention, an OECD conference on the misuse of science, as well as through organizations such as the InterAcademy Panel and the World Medical Association.[5] Following the first recommendation of the 2003 report *Biotechnology Research in an Age of Terrorism* that:

> national and international professional societies and related organizations and institutions create programs to educate scientists about the nature of dual use dilemma in biotechnology and their responsibilities to mitigate its risks[6]

the US National Science Advisory Board for Biosecurity was charged with developing 'mandatory programs for education and training in biosecurity issues for all scientists and laboratory workers at federally-funded institutions'. Various initiatives are underway in the United States to produce dual-use educational modules and materials.[7]

Despite the breadth of calls for further education and awareness raising of scientists, grounds can be offered for doubting the likely unanimity regarding specific initiatives. A basic question herein is one central to education in general: is the purpose to facilitate individuals' thinking about a particular issue or is it necessary to instill a new understanding? Stated somewhat differently, the tension is whether educational activities should attempt to further individual reasoning or whether it should seek to challenge certain ways of thinking in order to promote others. The former implies merely drawing further attention to dual-use issues in order to spark further deliberation, the latter that an authoritative understanding of the potential must be instilled and promulgated. As Billig and colleagues point out though, in practice efforts to 'bring out' and 'implant' knowledge in education are not opposed extremes, but rather the two tendencies are often mixed in complex ways.[8] Thus, crucial questions exist for any program of education regarding how it would handle disagreement and how it would be judged to have 'educated' scientists. At the time of writing in 2006, consensus on such matters is far from being agreed.

Moreover, unless one simply assumes that the facts of the matter and their proper interpretation are obvious, then any efforts to educate scientists are going to have to address the complicated concerns discussed in previous chapters regarding what counts as the relevant facts and what they should be taken to imply.

These points regarding 'education for what?' and 'education about what?' beg the question of 'education by whom?'. Even if one assumes self-governance of experts is appropriate, which scientists should be deemed the bearer of (the most) authoritative knowledge? If that community is itself divided on how to respond, then moving forward with some program of education would not be straightforward.[9]

Moving forward with social analysis

Questions about how to move forward with social analysis about dual-use parallel many of those already mentioned for the substantive topic itself. At a general level, it is relatively easy to identify possible functions for social research. For instance, following on from points made in previous chapters, it seems fair to make the generalization that the issues surrounding the security implications of research data and findings have not been subject to widespread discussion in the past by Western life science communities.[10] Yet, the extent to which researchers have considered dual-use issues and what they make of them has a direct bearing on what sort of educational programs or other initiatives would be prudent. The past dearth of empirical data on such matters hampers the creation of well-tailored responsive measures.[11] As another contribution to social research, Chapter 2 argued that there has been a lack of detailed public elaboration regarding how cases of dual-use research are assessed in practice. Research that sought to obtain a sense of individuals' potentially varied and nuanced reasoning would be another option. The understanding gained from this could inform educational and other activities.

On closer inspection though, even such general contributions raise a myriad of questions regarding exactly what should be done, its specific purpose, and what types of data should be sought. In addition, questions can be asked of the status of any research. Social analysts confront similar, if not more entrenched, questions about the status of their knowledge claims in comparison to life scientists. As elaborated previously, policy discussions of dual-use issues often frame them as turning on highly technical matters. This then justifies regarding scientific expertise as essential and downplaying the role for non-technical experts such as 'the public' or politicians. Social researchers, of course, can attempt to secure their own basis for expertise. One option would be seeking to represent the collective views of scientists through conducting survey or interview research. Any understanding formed, however, is a second order one in the sense that it is based on interpreting statements made by individuals who themselves are seeking to interpret issues. Unless the formation of this second-level understanding is regarded as a straightforward process of tallying replies to self-evident topics (and Chapter 2 suggested many reasons why this was not the case), thorny questions must be addressed about how claims are substantiated.

As another point of consideration, in inquiring about scientists' knowledge of dual-use matters or seeking to partake in educational initiatives, many questions could be raised about the role of analysts in setting the bounds of discussion. Who determines what the key questions are, when they have been satisfactorily addressed, and what can be inferred from these responses would be important issues here. If the potential for dispute regarding the relevant 'facts of the matter' complicates making decisions about dual-use issues, it also complicates identifying topics to examine as part of social research.

In this vein, consider some reflections on conventional interviewing techniques. In 2003 I conducted 16 semi-structured one-to-one interviews at British universities with life scientists investigating the functioning of muscarinic acetylcholine brain receptors.[12] These interviews sought to determine how scientists defined the possible bioweapons applications of their research and what they thought of ongoing debates about security oversight measures.[13]

In relation to the themes in this chapter, I can offer a few personal reflections on the interviews. First, it could be argued that they constituted a form of education in themselves because the interviewees indicated little prior awareness of bioweapon prohibitions or ongoing dual-use policy deliberations. Only three stated ever having considered the weapons applications of their work (two of whom only after being approached by US military establishments).

Second, the interview interactions repeatedly bordered on the awkward. As scientists were being asked about possible negative consequences of their work that they had largely hitherto set aside as well as the prospect of restrictions on their activities, the management of tension was a continuing preoccupation. In terms of having a mutual way dialogue, two possibilities had to be avoided. One, the conversation turning into an apparent 'grilling' session where I asked question after question to which the interviewees would often struggle to respond given their lack of prior reflection on the topics covered. Two, the conversation turning into a narrowly technical assessment of the feasibility of moving from scientific findings to malign applications for which I would struggle to do anything more than listen politely.

Another aspect of awkwardness was the proper extent of challenging. Many participants offered blanket reasons against any additional security controls on research by suggesting such measures would comprise the open exchange of science or that restrictions would be futile given the extent of knowledge already in circulation. In light of commercialization and competitiveness pressures noted in Chapter 2, for instance, it was readily possible to question the veracity of the former argument. In a situation of one-to-one interviewing, this sort of challenging threatened to degrade the interview into an adversarial confrontation. In this case that meant an opposition between a junior sociologist and (almost always) a more senior biologist about the implications of his work.

As a third reflection, and following on from those already made, going around and asking questions could not be seen as a neutral act. Raising hitherto largely ignored concerns about the negative potential of life science research and at least posing questions about the need for a response was consequential in the manner that heightened the importance of this topic among the competing demands for scientists' attention.

Consider the relevance of these personal reflections in relation to the somewhat cleaned up interview extract below with a leading British scientist in the study of brain receptors. The reference to 'the tap of science' in my question refers back to a previous comment he made that science should be seen as producing flowing outputs and any dual-use controls should, in effect, be thought of as taps restricting that flow:

Excerpt 3.1

BR: Let me ask you about, hmm. You mentioned these issues before about the tap of science, and how far you want to turn it off. And here I suppose one of the sort of rejoinders to the concerns that you mentioned that has been raised by others is about commercial links with industry. The argument is something like this: academic, publicly funded research, whatever it is, is in many ways already constrained in its circulation because of commercial relations and commercial considerations. So national security people would argue, OK, well we're not asking anything different than what's already been done for commercial reasons. And I wonder if you think that's an inappropriate parallel to make?

((21 second delay))

RES: That is a very difficult issue to address in terms of particular contexts, you know. As I say I'm ((long pause)) what you're essentially saying is that scientists ((long pause)) presumably within and without, and outside the academic community should be asked to sign the equivalent of the Official Secrets Act. Ah, including people, yes, where your rights to disseminate information are controlled by a governments or third party, a government's agency. If you're actually going to propose that then that would be an obligation on people approaching industry and not in industry, in academia, and in research institutions and that has good and bad consequences. You're getting towards Big Brother there.

BR: Yes, it's not. I realize in asking these questions it might sound like, I'm sort of advocating them

RES: ((Interjects)) No.

BR: this is the debate in the US ...

The words in double parentheses here are my description of what happened. Although at one level a friendly interview, it was one I would characterize as preoccupied with the factual assessment of the practical feasibility of moving from research to bioweapons. This framing meant that the 'conversation' was fairly one-sided in terms of who was speaking to the issues deemed relevant. The interviewee repeatedly downplayed any dual-use potential of his work through detailed scientific arguments.

Bringing up the question of the parallel between 'commercial' and 'security'-inspired controls acted to disrupt the general direction of the discussion by reframing the issues at stake. In making sense of this part of the exchange, it is perhaps worth noting that it was preceded by earlier discussions about the scientist's well-established links with the pharmaceutical industry. As a result, before the commercialization parallel was brought up, the presence and overall acceptance by the interviewee of links was already established. Presumably this made it difficult for him to reject the relevance of the parallel. In any case, as indicated by the exceptionally long pause before the response, this question dramatically interrupted the flow of the interview. Certainly I experienced the gap as awkward and uncomfortable, and judging by the interviewees' facial and body expressions it appeared he took it as a form of criticism. The subsequent attempt to distance myself from actually advocating the points being made was an attempt to smooth over any interactional awkwardness.

A pragmatic response to questions with learning and inquiry

Given the points of the last two sections, then what sort of social research should be carried out? What might be expected to be gained from this? The previous argument suggested addressing such questions requires more than a narrow examination of the pros and cons of possible methods, but rather a considered rationale for any work. As also suggested, learning should be a core component of any strategy; learning in the sense of both trying to understand the meaning and implications of dual-use research as well as questioning the basis for deriving any particular understanding of these matters. The latter requires a way of learning about learning.

How might this be achieved? The social philosopher John Dewey (1859–1952) spoke to many of these issues. One of his many concerns over his long and varied career was how inquiry might help in coping with 'problematic situations' where things did not 'hang together'. In other words, situations where indeterminacy, uncertainty, and doubt combined to caution against carrying on as usual, but frustrated determining what should be done. Inquiry here should start from problematic experiences rather than from abstract questions. 'Any philosophy', he wrote (and a statement that for Dewey could be applied to any type of investigation), 'that in its quest for certainty ignores the reality of the uncertain in the ongoing process of nature denies the conditions out of which it arises.'[14]

What Dewey and contemporary pragmatists seek from inquiry is a way of practical problem solving. So, rather than engaging in a search for certain universal truths, the goal is to make situations manageable in order to find a way of improving them. Given the complications, complexities, and changing considerations associated with many issues, expecting to fix the world down into neat cause and effect relations between given variables is unrealistic and even misguided. As such, a rather modest notion of truth is sought. Herein to 'establish a truth pragmatically is to settle a controversial or complex issue for the time being, until something comes along to dislodge the comfort and reassurance that has thereby been achieved, forcing inquiry to begin again'.[15] Stated differently, what is regarded as 'true' is not established once and for all time, but rather contingent on the adequacy of the understanding formed for particular human purposes. Claims which aid such purposes, which make a difference in our understanding of the world, have a 'warranted assertability'.

For pragmatists the establishment processes of experimental inquiry that can aid in learning are essential. What is also essential is attention to the fallibility and purposefulness of any methods employed. Those methods will necessarily place in the background some points in favor of scrutinizing others. Cochran, quoting Dewey, outlined the generic traits of experimental inquiry as the following:

> Firstly, 'all experimentation involves overt doing, the making of definite changes in the environment or in our relation with it'; secondly, all experiments are 'directed by ideas which have to meet the conditions set by the need of the problem inducing the active inquiry'; and finally, 'the outcomes of the directed activity is the construction of a new empirical situation in which objects are differently related to one another, and such that the consequences of the directed operations form the objects that have the property of being known.'[16]

From these points it follows that inquiry should be seen as a practical, intellectual, action-orientated process that is tightly coupled to the problematic matters under question and the outcomes sought. Inquiry does not so much become finished at some stage because the truth is found as the need for it is reduced because of the temporary resolution of problematic doubt.[17]

As such, a good deal of emphasis in pragmatism is placed on getting beyond conventional dichotomies and binaries. The distinction between 'action' and 'research' is one. So, too, is any stark distinction between those that initiate research and those whom the research is 'about'. The word 'inquiry' itself (as opposed to 'research') signals a greater inclusiveness. In relation to the case of dual-use research, engaging 'scientific' and 'social' analysts alike in a process would be essential. Herein, in the service of finding out 'what works', the relevance of the distinction between social and scientific researchers (so ubiquitous in previous chapters) should dissolve away. Also suspicious for pragmatists are distinctions between description and evaluation, objectivism and constructivism, as well as 'commonsense' knowledge and 'scientific' knowledge. Again, such lines throw up unnecessary obstacles for engaging in practical problem solving.

In combination with the arguments given in previous chapters, the account provided above suggests a variety of reasons why a Deweyan-type rationale might be useful in the examination of dual-use issues:

- This places learning and experimentation at the center of practical-inquiry.
- It acknowledges research as a consequential form of intervention.
- The aspiration to transform social relations through improving situations sets an ambitious yet achievable goal.
- It stresses the importance of tightly coupling empirical agendas with an understanding of the problematic substantive matters under examination.

Rorty: less method, more conversation

To say that a Dewey-type rationale speaks to many of the concerns raised earlier about dual-use research only begins to provide an initial starting point for setting out what should to be done by way of inquiry. This can be illustrated by considering the contrasting directions subsequent individuals have taken his work.

Richard Rorty is perhaps one of the best known contemporary academic disciples of Dewey. As a philosopher, much of his dialogue with Dewey centers on the aforementioned theme of treating knowledge as a means of problem solving rather than truth grasping. As Rorty advocates: 'The purpose of inquiry is to achieve agreement among human beings about what to do, to bring about consensus on the ends to be achieved and the means to achieve those ends.'[18] This applies to all types of inquiry, be that research into the physical properties of materials or ethical deliberations about everyday conduct. Achieving such co-ordination should not be seen as a matter of agreeing 'The Truth' in some absolute sense, but rather agreeing on what counts as useful knowledge for a certain purpose. One of the central aims of this line of pragmatic thinking is to shift away from the sorts of long-standing concerns in philosophy about what is objective or subjective, real or apparent, found or made. Rorty simply asks that we turn our attention elsewhere.

One such course is finding more and better vocabularies for talking about issues of concern. As he says of his tradition of philosophy, 'the point ... is to keep the conversation going rather than to find objective truth.'[19] Herein:

> To see keeping the conversation going as a sufficient aim of philosophy, to see wisdom as consisting in the ability to sustain a conversation, is to see human beings as generators of new descriptions rather than beings one hopes to be able to describe accurately.[20]

In generating new descriptions it is possible to ward off premature closure of issues, encourage new forms of thinking, and come to more useful (though no doubt transitory) understandings. Stated in somewhat different terms, 'One should stop worrying about whether what one believes is well grounded and start worrying about whether one has been imaginative enough to think up interesting alternatives to one's present beliefs.'[21]

But just as Rorty is not preoccupied with how 'objective truth' might be established, so too is he not concerned with specifying some general method for deriving the truth. The idea that 'science' has a method by which its claims are somehow set apart from other bodies of knowledge is regarded as an antiquated understanding. Following on from this desire not to set fixed criteria, just what counts as a 'good' or 'useful' conversation is something that itself should emerge through problem solving rather than being preordained on the basis of existing beliefs or theories. Rorty's reading of Dewey is one that sees him as generally resistant to establishing fixed methods for inquiry,[22] an interpretation not shared by everyone.[23]

Argyris: more method, better conversation

One of those who extends Dewey's thinking along more method-based, prescriptive lines is Chris Argyris. During almost 50 years of work, he has sought to document and address the following 'problematic situation': despite widely professed commitments, many organizations and inter-personal relations are characterized by features that discourage inquiry and learning.

Argyris identifies the basic source of anti-learning tendencies as a defensive mindset. This mindset stresses four basic values:

1. to remain in unilateral control;
2. to maximize 'winning' and minimize 'losing';
3. to suppress negative feelings; and
4. to be as 'rational' as possible – by which people mean [unilaterally] defining clear objectives and evaluating their behavior in terms of whether or not they have achieved them.[24]

He argues that such '[d]efensive reasoning is omnipresent and powerful'[25] as it can be found across cultures and at all levels of organizations. Attempt to control situations and avoid oneself or others being threatened means there is little testing of the basis for views and evaluations. Defensive reasoning leads to the use of covert attributions of motives, scapegoating, the treatment of one's own views as obvious and valid, and the use of unsupported evaluations. The end result is the reproduction of invalid assessments and inferences through self-reinforcing and self-sealing routines that inhibit inquiry.

A central distinction in Argyris's work is between the many theories espoused by individuals about how they interact and the near universal way (defensive) interaction takes place. This disjuncture has substantial implications for the ability of individuals to learn from past experience. He contrasts so-called Model I forms of learning, indebted to defensive reasoning, and Model II form which is in-line with what is often espoused and more desirable:[26]

Model I	Model II
• Offer un-substantiated attributions and evaluations	• Ensure reasoning is explicit and publicity test or agreement at each inferential stage
• Unilaterally assert evaluations	• Advocate position in combination with inquiry and public reflection
• Make covert attributions	• Publicly reflect on reactions and errors
• Protect inquiry from critical examination	• Inquire into your impact in learning

While the sort of defensive reasoning that characterizes Model I might be useful in accomplishing certain narrow goals, it limits the extent of learning. Argyris stresses the importance of making individuals aware of their Model I responses in order to begin to move towards more robust forms of inquiry.

To this end, he has devised techniques for both illustrating the problems of defensive reasoning and fostering individuals' awareness. One is the relatively simply exercise of getting people to identify a key issue within their organization and to write down actual or hypothetical conversations they have had about that issue. In addition, the individuals are asked to also note thoughts or feelings left unsaid in such exchanges. The following is one such account:[27]

Case illustration: the easing-in CEO

Thoughts and feelings unsaid	The conversation
I hope we will work co-operatively. I fear we won't.	CEO: I'm sure that you and I share the same goals. We need to rethink out latest cost estimates. Other: The latest estimates are not the most reliable ...
What's wrong with him? He's missing the most important point.	CEO: I can see some costs can be reduced (by your recommendation), but that still leaves us with a very large problem. Other: The original estimates were produced by others. We never really agreed to them.
He doesn't want to accept ownership; he wants to make me responsible.	CEO: We will have to use these estimates. That's the reality.

Although in this case the CEO of a company had various concerns and made various attributions about another person, these were not voiced and as a result remained untested. As Argyris states, these conversational practices were carried out even though in this case the CEO had advocated direct talk within the company. Rather than speaking in an open manner which encourages further dialogue, here the CEO is 'easing-in'.

Easing-in meaning the practice of hinting at concerns through asking particular questions that the listener is supposed to recognize as critical without an explicit statement of that evaluation. Furthermore, though, it seems reasonable to presume that the other party to a dialogue would be making their own evaluations and attributions. Because both are likely to suspect that each other is acting in such a manner, that they are doing so itself has to not be mentioned. Thus, a self-reinforcing process is enacted whereby concerns are left unexamined.

Another example of how defensive reasoning can reduce the scope of interaction and therefore learning was given in pages 61–2. That recounted an awkward part of an interview between a scientist and myself. In this I offered a parallel regarding the similarity between national security and commercial-inspired research controls. Whatever thoughts and feelings were generated in the exchange were left largely unstated and unexamined. Instead, deflection, silence, and hesitation were notable features of the exchange. While it is possible for me to speculate about what the interviewee was 'really' thinking (and vice-versa), these are just the sort of Model I-type strategies of attribution that Argyris criticizes.

One of the troubling aspects of defensive routines is that even when individuals are made aware of them and undertake to correct their behavior, they consistently act in a manner in line with the types of practices they are seeking to correct.[28] Argyris illustrates this through recounting reactions to the type of two column technique.[29] So in one case, a group of financial executives were asked to produce written mock conversations.[30] Included among the thoughts and feeling left unsaid in their write-ups were statements such as: 'He is clearly being defensive', 'He is baiting me now', 'Will he never be able to change?', and the like. When asked to comment on what conclusions participants made about the individuals making these statements (in other words, one another), group members responded by giving various negative evaluations such as the individuals were 'opinionated', 'frustrated', 'angry', 'not listening', and so on. Yet these responses themselves were based on the same kind of unsubstantiated attributions and evaluations the participants were criticizing in those that made the statements. It is this sort of practice which led Argyris to conclude that 'There appears to be a systematic discrepancy between the writers' expressed aspirations to learn and to help others to learn and their actual behavior, which is largely counterproductive for learning. The individuals are systematically unaware of the ways in which they produce their own unawareness.'[31]

Scholarly research in the social sciences, management, and other fields is likewise said to embody anti-learning strategies. So Argyris critiques the standard 'norms' of academic inquiry: seeking to describe the world 'as is' rather than finding ways of improving it, failing to help those being researched to develop the practical skills necessary for improving learning, and downgrading the relevance of positive normative goals. Model I reasoning in most social science research means, 'It is the researchers who are in unilateral control over the methods; who define the criteria of victory; and who design research conditions that minimize the expression of feelings, especially of doubt, bewilderment, and at times anger being dominated by researchers in the name of validity.'[32]

What follows from the points above is that, in line with other pragmatist thinkers such as Dewey, Argyris agrees on the need for inquiry to be an inclusive process of practical action that strives for change. In contrast to Rorty's call for open-endedness regarding how this is done so as not constrain future possibilities, for Argyris facilitating further inquiry requires ways of disrupting taken-for-granted practices. Knowing how best to intervene in turn requires a scientific approach. His type of research, given the name 'Action Science', seeks to offer such a scientific basis for inquiry. It does so by putting the public testing of empirically disconfirmable claims at its center. Yet, within this call for 'Action Science' there is also a need to reconsider traditional notions of what counts as scientific. So Argyris criticizes forms of social research which strive to mimic artificial experimental conditions in the physical sciences. Instead, he advocates undertaking research which through iterative processes of action and change enables the greatest reflection on the substantive concerns of individuals and the norms of inquiry.

In support of combining action and science, one way of countering Model I forms of learning is the seemingly simple suggestion of promoting interactions wherein the data, inferences, and assessments are made explicit so as to be subjected to public testing. Interactions should be conducted in such a way as to test the 'ladder of inference' underlying statements. This prescription is not just one of getting individuals to overtly challenge others' claims, but to conduct dialogue in a way that fosters further inquiry into the basis for claims. Such techniques can play a part in the wide-ranging organizational and personal changes Argyris identifies as preconditions for moving beyond defensive thinking. So in relation to the commercialization exchange given in the previous section, instead of simply distancing myself from advocating the security–commercialization parallel by suggesting it was being made by others 'out-there', it would have been possible to ask the interviewee what lead him to conclude that I thought scientists should sign the Official Secrets Act. It also would have been possible for me to more thoroughly elaborate the reasoning for my choice of asking the question and to test whether the interviewee thought this was appropriate.

Contrasting pragmatics

Previous chapters argued that both devising policy measures and conducting research into dual-use issues poses a variety of challenges and choices. Starting from a sense of the scope for disagreement and uncertainty regarding the question of 'what should be done?', this chapter has considered the role of learning in devising responses. So, raising awareness and educating scientists about dual-use issues has been a central plank of (self-governance) recommendations coming from science policy circles. Yet, as argued, such a call itself raises many difficult questions about the form and purpose of any activities.

This chapter has moved on from such conclusions to question the place and purpose of social research. As argued, this research should be understood as a particular way of intervening into existing relations. Just what is asked and how are key questions. The initial case was made for problem-orientated research that aims both to collect information about the reasoning of practicing scientists and to engage with them in a process geared towards mutual learning.

By way of moving forward with both the substantive and methodological concerns, this chapter proposed that the pragmatic thinking associated with John Dewey provides a useful starting rationale for approaching the question of 'what should be done?'. His emphasis on experimentation, practical intervention, mutual learning, and linking the form of inquiry to an understanding of a 'problematic situation' is in line with points raised throughout previous chapters. However, as suggested by the divergent manner in which Dewey has been appropriated by prominent individuals such as Rorty and Argyris, moving from such emphases to specific interventions raises its own challenges and choices. The need for and possibility of a 'scientific' method for inquiry was one source of alternative reasoning. While Rorty eschews talk of method in favor of attending to creativity, for Argyris robust creativity can only be attained through methods that break down entrenched routines. Part II of this book moves from the general points outlined here to consider a specific program of intervention undertaken by the author.

Part II

The Conversations – Themes and Actions

4
Learning to Respond

In many respects, the previous chapters have functioned as a ground-clearing exercise for thinking about what should be done by the way of empirical inquiry. Part II now describes a line of inquiry undertaken with bioscientists by Malcolm Dando (University of Bradford) and the author. In the analysis of these exchanges, as in Part I, the substantive challenges and choices of specifying 'the problem' with dual-use research will be discussed alongside the challenges and choices of conducting inquiry into this issue. As in Part I, the substantive understandings expressed and the means of inquiry will be treated as inter-related. The importance of this will be readily apparent because what follows is not just an analysis of how other individuals have defined the issues at stake. Instead, it is a personal account of an effort to initiate and sustain a conversation.

This chapter begins this agenda in two ways. First, it moves from the general research rationale outlined in Chapter 3 to a specific form of inquiry. In the previous chapter, the pragmatic case was made for a problem-orientated form of inquiry that considers the reasoning of practicing scientists regarding matters of dual use while striving for mutual learning. One way of seeking this is detailed here. As a second contribution, this chapter provides a broad overview of the types of exchanges that took place in these dialogues. In examining the prominent themes discussed, a number of issues mentioned in earlier chapters are scrutinized, including the responsibilities of scientists, the characteristics of science as an activity, and the relation between technical experts and others. The other chapters in Part II develop the initial analysis provided in this one in greater detail.

What to do?

How then might inquiry be undertaken so as to promote mutual learning? A consideration of method is one place to begin to address this question. It was suggested in Chapter 3 that the form of any inquiry should be linked with a working understanding of the substantive matter in question. In relation to dual-use implications of life science research findings and techniques, an initial working hypothesis stemming from Part I was that the relative lack of past professional attention to these issues meant it was unlikely that many bioscientists would have well-developed thinking in this area. This, coupled with the dilemmas associated with determining how to respond, suggested it would be sensible to explore individuals' thinking in some depth. The searching and questioning involved should extend to the matter of what questions should be asked.

As argued also in the previous chapter, standard one-to-one semi-structured interviews might allow for some mutual learning and reflection. As a method for engagement and dialogue, however, this technique had certain limits. It was contended that the promotion of two-way dialogue was hindered in such interviews by asymmetries in knowledge about both specialist scientific areas and policy developments. Further, the potential for dual issues to be personally and professionally threatening meant that standard one-to-one semi-structured interviews had additional drawbacks.

Focus group method

What type of method for inquiry might be more appropriate? In recent years, much has been written about the potential of focus groups to promote deliberation.[1] 'Focus groups' differ considerably in terms of their makeup. In the main though, they generally consist of assembling 5–9 people together to collectively discuss a predetermined set of issues under the guidance of a moderator.[2] Two advantages are frequently claimed for such group interactions. First, they are 'ideal for exploring people's experiences, opinions, wishes and concerns. The methodology is particularly useful for allowing participants to generate their own questions, frames and concepts and to pursue their own priorities in their own terms, in their own vocabulary.'[3] As such, focus groups enable a rich examination of the *whys* behind individuals' thinking. As a second advantage, they entail 'the explicit use of the group interaction to produce data and insights that would be less accessible without the interaction found in groups'.[4] As Krueger states, 'focus group interviews produce data derived from a group process in a focused manner. As a result, participants influence each other, opinions change, and new insights emerge. Focus group participants learn from each other, and things learned can shape attitudes and opinions. The discussion is evolutionary, building on previous comments and points of view.'[5]

In relation to exploring dual-use aspects of life science research, such advantages are desirable for a variety of reasons. As the security dimension of biological research results is (in certain respects) a rather novel topic for bioscientists (Chapter 1), understanding how they conceive of and frame the basic issues at stake is vital. Potentially at least, the interaction *between* peers could be a way of minimizing asymmetries in expertise and oppositional relations between those asking and those answering questions. As moderators do not have to press the same individuals again and again, group interviews allow for a space for personal reflection. Indeed, some have suggested focus groups might be particularly useful for exploring 'sensitive' issues.[6] The interactive dimensions can, in turn, serve educational purposes and foster change in people's thinking.[7] In this manner, focus groups are in line with pragmatist-inspired models for education which see interactions as 'places where a variety of voices can be heard and considered with care'.[8]

The said benefits afforded by this general method have, though, also been matters of criticism for some. The typical use of 'purposive' sampling and the often low number of sessions undertaken means those employing it rarely strive for 'statistical representativeness'.[9] In areas such as marketing, where focus groups have become common, they often serve the purpose of informing other kinds of research. When focus groups are held with those with pre-existing relations, then their interaction can be said to be 'contaminated'. Even if members are unfamiliar with each other, the group dynamic is said to result in conformity and individual censorship.[10] On a more practical level, the resources, expertise, and planning required for successful sessions are said to nullify many of the practical advantages of convening group-type interviews.

In response, proponents of focus groups have acknowledged that they often lack statistical generalizability and incur significant resource demands, but defended them by arguing that when properly done, on their own they can produce verifiable results.[11] It is often acknowledged that group interview settings can both produce conformity and encourage openness.[12] This gives reason to think individual responses offered in peer group dialogues are likely to differ in some respects from those given in one-to-one settings. Crucially though, this does not thereby imply the latter should be regarded as more authentic.[13] With any type of social research dangers exist that individuals will offer socially preferred responses and rationalized justifications.[14] Each method should be scrutinized in terms of its underlying assumptions and the trade-offs in the commitments made. For the purposes of the issues under consideration here, since what was needed in the case of dual-use issues was an exploratory process of peer engagement, group session methods had definite overall advantages.

As an additional point of concern, interactive group discussions are not straightforward to analyze. Their *interactive* dimension means that individual sessions can evolve along unique lines. This makes it problematic to provide cross-sessional figures about what was said.

Finally, more fundamental questions can be asked about the status of the responses given. Morgan, as with many others cited above, maintains that focus groups are a way of getting closer to 'participants' experiences and perspectives'.[15] Yet, in-line with the comments made in Chapter 3 about language as action, much of the recent work in social science regarding the discursive status of accounts would counsel against treating accounts made in some particular form of interaction as simply representing individuals' attitudes.[16]

Taking this orientation forward in the study of environmental risks, for instance, Waterton and Wynne critiqued the idea that attitudes on many complex topics should be regarded as stable, coherent, and unambiguous entities that can be tapped through surveys or interviews. Instead, attitudes expressed are done so: '(a) in relation to their relevant social context ...; (b) interactively – that is, they actively form attitudes though the opportunity of discussing issues that are not often addressed; ... and (c) as a process of negotiation of trust between themselves as participants and ... researchers'.[17] In the case of risks assessments, that might mean that responses given (be they as part of surveys, one-to-one interviews, or group interviews) depend on matters such as: the historical context of concern for particular participants, the sequence of what questions and responses have already been made, the perceived uses of the research, trust in institutions that control risks and pose questions, and the sense of agency of respondents. A general implication of this and related studies is the inappropriateness of treating responses made about complex topics as discrete entities that should be added together to provide a summation of individuals' 'attitudes'.

Some lingering questions

Despite the burgeoning literature about focus groups, arguably it fails to address crucial issues surrounding the openness of, and rationale behind, moderator questioning. In this respect, Morgan advises that '... focus groups allow you both to direct the conversation towards topics that you want to investigate and to follow new ideas as they arise.'[18] Yet how 'directing' versus 'following' should be reconciled in practice is rarely a topic of detailed consideration even when noted as a crucial. So Kitzinger advocates that 'trying to maximize interaction between participants could lead to a more interventionist style: urging debate to continue beyond the stage it might otherwise have ended, challenging people's taken for granted reality and encouraging them to discuss the inconsistencies between participants and within their own thinking',[19] but without any further elaboration of the rationales for and implications of intervention strategies. The tensions with questioning not only have implications for assessments of 'rigor' of focus groups but also for the claim by Kitzinger and Barbour that this method allows participants to 'generate their own questions, frames, and concepts, and to pursue their own priorities in their own terms, in their own vocabulary'.[20] The standard advice given that facilitators should be non-judgmental, impartial, and non-expert further complicates knowing how and what interventions could be undertaken.

The logic for changing planned questions is another area of concern in focus groups. In an effort to ensure the rigor of results, maintaining consistency in the questions asked has been identified as vital.[21] Revisions can be justified if new issues emerge or certain questions repeatedly generate predictable and unenlightening discussions, but the need to draw overall findings stands against the desirability of revisions once the research process has begun.

To the varied issues associated with how facilitators should conduct themselves, the pragmatist lines of thinking discussed in the previous chapters would counsel the importance of an experimental approach. Striving to 'box in' interactions to achieve consistency and control risks bypassing opportunities for learning. In emphasizing experimentation, repetitious responses should not only be taken as reason for ceasing inquiry, but as a reason for intensifying it. For instance, Morgan advocates that if 'the discussions reach saturation and become repetitive after two or three groups, there is little to be gained by doing more',[22] and furthermore that if one 'can clearly anticipate what will be said in the next group then the research is done'. Instead of assuming this orientation, which is indebted to treating inquiry as a research process of gathering information, the emergence of common themes can be treated as an opportunity to find ways of examining the presumptions and inferences informing shared thinking.

Who to have a conversation with?

In seeking a conversation with life scientists through a focus group-type method, just who should partake in that dialogue is of central concern. In the range of all those in industry, government departments, and educational institutions who undertake life science research, our dialogue was with those in university life science departments; this including faulty, technical support staff, postgraduate students, and to a lesser extent undergraduates.[23]

These seminars took place within universities in the United Kingdom: 76 universities with biology research seminar series were approached and 26 seminars were held in total (two being pilots) involving 624 participants and lasting between one and two and half hours. In total 13 seminars were held in England (excluding Greater London); six in universities in Greater London, three in Scotland, two in Wales, and one in Northern Ireland.[24] Another was held in Germany to see if any marked differences were noticeable. Approaches were made to offer a seminar either through contacting the individuals in charge of department seminars, or, when possible, through personal contacts within the departments. In contrast to these fairly wide-ranging US initiatives discussed in Chapter 1, activities undertaken in the United Kingdom have been more limited (see Chapter 1). Nothing like the scale of funding increases for biodefense or the intensity of life science–national security dialogue had taken place prior to 2005. Much of the policy discussion about the malign applications of research in the United Kingdom has centered on community self-governance measures such as a professional code of conduct and increasing university ethics teaching provisions.

There were a number of reasons for this sampling: One, many of the novel dual-use controls being proposed are primarily designed for civilian research outside of government, military, or corporate laboratories. Two, as university members are already subject to numerous biosafety regulations and research protocols, participants would have been familiar with general issues of research governance. Three, universities are relatively open institutions (for instance, in comparison to industry) that have a tradition of facilitating discussion about societal issues. Fourth, in terms of encouraging further/future deliberation, the educational role of universities makes them a highly appropriate site.

The focus group-type discussions were given as part of the pre-existing university seminar series. The use of such series had a number of practical benefits: the room and equipment was already arranged and no monetary compensation was required as in many focus groups (and therefore its impact was not relevant). Because of expectations in British universities regarding attendance, this forum was a comparatively straightforward one to secure audiences with varied profiles who were also relatively at ease with the setting location. As another negative, though, the lack of control over the specific venue location meant the quality of audio recordings suffered.[25]

This forum had mixed advantages and drawbacks in relation to the size of the groups. University departments differ considerably in terms of their size and composition. The number of those present ranged from 5 to 75 with an average of 24. No noticeable differences were apparent in the ease of initiating and carrying on discussions due to audience size. While this average size enabled many people to be involved, it was also significantly higher than typical focus groups. As such the seminars had to trade off between the space it enabled for individuals to respond and the number of those attending. In the end, the lack of familiarity of attendees with the issues raised – and therefore the typical exploratory discussions – fitted relatively large gatherings.

Likewise, it was recognized from the start that using pre-established groups had certain pros and cons. On the positive side, in such groups individuals can question each other on the basis of mutual experience and knowledge. They can also provide a basis for continuing post-seminar peer discussions. Since many of the issues discussed related to how particular institutions might govern research, conducting discussions within existing department seminar series had certain justification. On the negative side, any pre-existing relations of authority threatened to produce conformity to the views of certain individuals.

Dialogue and 'real time' experimentation

While focus groups are generally in line with the goals of inquiry identified in Part I, outstanding issues remain about their most appropriate form. In the university seminars Dando and I conducted, we departed from many of the standard prescriptions for conducting focus groups because of the importance attributed to learning.

In line with a general prescription by Argyris, the main rationale for this was to investigate the reasoning informing individuals' appraisals of issues by making the data, assumptions, and inferences underlying responses explicit in order to publicly test them.[26] In order to achieve this progressive inspection, the *transformation* of questioning on the basis of experience was essential. So, in each seminar, certain key topics were covered: Are there experiments or lines of research that should not be done? Is some research better left unpublished or otherwise restricted in its dissemination? Are the envisioned systems of pre-project research oversight strategies sensible? Yet, the seminars differed in the ordering, content, and emphasis of the particular form of questioning. To do this we structured the seminars such that *within* particular sessions we could question the basis for previously stated evaluations by revising the seminars *between* sessions.

Following on from the analysis given in Part I, at the start of the seminars in late 2004 we had a number of working premises and priorities to test out:

- As central policy concerns included whether certain research should be conducted, published, or subjected to oversight measures, we needed to discuss these concerns. As much of the literature in the social study of science suggests there are many choices in what is published, carried out, or controlled, *how* these acts are undertaken also had to be part of the discussion.
- Whether any oversight or other controls would be seen to jeopardize the norms or ethos of science.
- The relevance of mainly US-centered biosecurity debates for British-based bioscientists.
- In line with the dominant framing of many policy analyses, where the proper 'balance' between security with scientific freedom rested.
- With the attention in the United Kingdom given to 'codes of conduct' as a possible response by the scientific community, what might be in such a code.
- Whether participants had a relatively low level of knowledge of dual-use policy debates.

These served as the initial basis of questioning that would be subject to later modification as we learned how to question. This responsive orientation was recognized as important from the start because before going into the seminars we were not at all sure whether the sort of dialogic format envisioned would be feasible. The fear was that the novelty of the seminar proposed – in contrast to most scientific seminars – meant that we would be forced to make radical revisions simply to get the audience engaged.

The remainder of this chapter provides a broad overview of the discussions in the seminars, one that attends to their ongoing transformation on the basis of the experience gained. The analysis presented was largely formed during the period the seminars were conducted.

Later chapters in Part II of this book provide an analysis conducted after the seminars were completed.

Introductions

The seminars typically began with self-introductions by Dando and me, a brief mention about the topic of dual-use research and our assessment of the importance of initiating discussion about this with practicing researchers, and a request for permission to make anonymous audio recordings.

The seminars differed in important respects from common prescriptions for focus groups. As focus groups often try to 'tap' individuals' experiences or preferences, the advice is often given to start with general, bland, and non-challenging questions that can loosen up participants for more intense questioning. However, given our initial presumptions (later confirmed) about the lack of consideration or awareness of dual-use issues among practicing researchers, operating in this manner both had less justification and risked losing the attention of the audience. Instead, immediately following the introductory remarks, one of the controversial dual specific cases was described and the question asked about what should be done (i.e., either the interleukin-4 mousepox experiment that inadvertently suggested a way to manipulate smallpox and the question of whether it should have been published or the artificial chemical synthesis of poliovirus and the question of whether it should have been conducted in the first place – see page 18).

An example of the sequence of slides and key questions from the first post-pilot seminar is given in Box 4.1. As indicated from this Box, the seminar consisted of a series of slides with information regarding the possible relation between current biomedical and bioscientific research and new weapons possibilities as well as the range of national and international measures currently being implemented or proposed. Discussions were initiated through fairly open-ended questions posed after an extended monologue regarding each slide.

Seminars 1–5

After the pilots, 24 seminars were undertaken. Despite our initial fears about the basic viability of the format and topic of the seminar, it proved a workable model for engagement. A number of noteworthy substantive themes emerged from the beginning which would be echoed in later seminars.

To the question 'Are there experiments that should not be done?', the vast majority of responses given supported undertaking the 'contentious'

Box 4.1 Initial slide titles, details, and questions

1. Title of 'The Life Science, Biosecurity, and Dual-Use Research' seminars
2. What are we doing?
   An explanation of the scope and goal of our research and seminars
3. Cause for Concern?: Synthetic Polio Virus
   Question: Should it have been done?
4. Cause for Concern?
   Slide detailing rapid advances in synthesizing capabilities since 2002 publication of poliovirus paper
   Question: Is artificial synthesis still a good idea?
5. IL-4 Mousepox Experiment
   Question: Should such experimental results have been widely circulated?
6. The British Reserve
   Slide presenting suggestion that British researchers had worked on IL-4 insertion in pox viruses in the late 1990s and become aware of its potential implications, but then only raised attention to this to public officials
   Question: How should researchers make their research results available to others?
7. Responding to Bioweapons Threats: Keeping Ahead Through Research
   Slide noting US biodefense activities that raised questions about their offensive potential and hence the appropriateness of trying to stay ahead of biothreats through scientific work
   Question: Should we always seek to 'run faster'?
8. US NAS Fink Committee
   Slide detailing proposed US system for the oversight of research
   Question: Would such a system be helpful or dangerous?
9. Spanish Flu: What Should be Done?
   Slide detailing efforts to reconstruct the1918 Spanish Flu
   Question: Are there any limits on what should be done or how it is communicated?
10. Data Access and Genomics Research
    Slide drawing on social studies of science that question extent of free communication in research
    Question: In practice does science work according to free and open communication?
11. Codes of Conduct
    Background information about British and international codes activities
    Question: What individual and collective responsibilities should be included?
12. Thanks and contact information

experiments cited – this was because the experiments enabled devising countermeasures and furthered benign lines of investigation. Also the possibility of the malign use of knowledge was said by some to be present in any research. Yet the most often given response for not refraining from certain experiments was that research results were in some sense inevitable. Herein, the question of *whether* something should be done missed the point that it *would* be done by someone sufficiently skilled. There were variations on the general theme of inevitability, with some saying that efforts to restrict research in only certain locations (e.g., universities, the West) would be futile, whereas others said that attempts to limit any particular experiment would be pointless because the underlying knowledge in the field would suggest novel malign applications of concern. Yet, whatever the variations, the common upshot was that any limitations or controls would be in vain.

Much of the policy debate about dual-use research is pitched in terms of where the balance should be struck between security and openness. However, this inevitability line of reasoning effectively rendered the matter of where the balance should be struck irrelevant. In this regard, we included a slide about the 2002 artificial synthesis of polio virus by Wimmer and colleagues (which we expected few researchers would say should not have been done). This was followed up by a slide indicating the substantial pace with which synthesizing capabilities have moved on since 3003 (as represented by a 2003 article by Craig Venter and colleagues[27]) to see if such information gave any reasons for pause. Slide 4 included a reference to newspaper reports that Venter had labeled Wimmer's virus synthesis work as 'irresponsible'.[28] One not untypical exchange about Slide 4 went as follows:

Excerpt 4.1

BR: ... the question now with this sort of slide is, well, does the pace of this technology suggested by this slide, does that change anyone's views about whether or not making viruses synthetically is a problem?

P2: It changes credibility, I mean somebody should have said that's, it's purely irresponsible 'cos they beat me to it, I mean that to me is academic sour grapes, somebody's beat him to it, that's the way I would read it, and then the guy goes and does it a year later. I think in a lot of these things the genie's out of the bottle, you can't unlearn something, once somebody has put even a proposal for postulation forward about how a process is going to happen it will be followed up, it's in the wider scientific community, somebody will pick it up. A little piece of my research work, the last line in the paper, I was doing some work on *** toxins, the last line in the paper *** spawned the little bit of work that I'm doing now, just the last line, a throwaway comment, I didn't know at that stage whether it's possible to synthesize *** toxin from *** bacteria and that's spawned my bit of research work, anything said in any of the scientific papers to the wider community is going to be picked up on, the genie is out of the bottle.

MD: So I think what I hear you saying is that even the scientific community, there are certain things which the scientific community is ready to do, it's got the technology, the ideas are in various peoples' heads and this thing rolls and stopping it here wouldn't matter very much because the idea would occur somewhere else?

P2: Yeah.

MD: And it would proceed anyway, so the argument that some people put about synthetic polio, they were worried that it might lead to Ebola and they would argue that their, their worries were enhanced by the speed with which Venter was able to do it doesn't really hold water because it was going to happen anyway.

P2: It was going to happen. But it's the control bit, morally, if you're talking about taking the moral standpoint perhaps it is wrong to do this type of thing, but it's almost like trying to stop the sun rising tomorrow morning, once it's out it's out.

MD: And have a certain level of technology and a certain development of ideas the sun's going to come out.

P1: But as you say, I agree exactly with what ((P2)) said about the genie being out of the bottle and everything, it's not so much the future problems, it's what is there already, particularly in the form of the Russian states ...

(Seminar 4)

Similar contentions of inevitability were made more than once in each of the first five seminars. Because science was so often characterized in this way, concerns about the pace of change or the topic of research (e.g., the reconstruction of Spanish Flu) would not affect assessments. Consequently, after this group of seminars, Slides 4 and 9 in Box 4.1

were relegated to possible probes that might be brought up should the discussion warrant it.

As to the issue of whether some research is better left unpublished or otherwise restricted in dissemination, the advisability of this was overwhelmingly doubted. Reasons given included the importance of communication in countering the deliberate and natural spread of disease, the limitations of the details in articles to enable the replication of research, and the status of publications as just one way researchers share information. A repeated assumption underlining many such appraisals was that government agencies would follow up on any threats identified in the literature. However, the reasoning informing evaluations were often quite complex and negotiated (see Chapter 6).

We wanted to question the basis for any all-or-nothing, yes or no formulations offered as part of our discussions about whether dual-use research results should be published. We did so by discussing the appropriateness of an example from the United Kingdom brought to the author's attention by an official where a group of unidentified researchers were said to have conducted similar experiments with IL-4 in the 1990s (slide #6 in Box 4.1). In their scientific publication, though, they did not highlight the inadvertent dual-use implications. Instead, they quietly informed government officials within the Health and Safety Executive of the possible malign implications. In contrast to the general points whether to publish, participants expressed a wide range of views about *how* results should be communicated after this alternative model for communication was proposed.

To the issue of whether the NAS Fink Report system of pre-project research oversight strategies was sensible (see Recommendations 2 and 4 in Box 1.1), repeated skepticism was expressed about its advisability. Reasons included the infeasibility of weighing risks and benefits, the possibility for this sort of oversight to broaden out once introduced, the ease for those with malevolent intent to circumvent controls, as well as the number of existing regulations already encumbering science. Overall, a strong recurring theme expressed was that since British universities were not the sort of place bioterrorists were likely to work, the dual-use concerns and controls were misplaced. What instead emerged as a crucial response for participants was staying ahead of potential threats that might arise through conducting further research.

As another point of note, while we originally thought many would not have been familiar with dual-use debates, the degree of unfamiliarly expressed was much more than anticipated. Based on the extent of coverage of dual-use issues in journals such as *Science, Nature, New Scientist,*

PNAS, and so on, we assumed that many participants would have at least heard of experiments such as IL-4 in mousepox as well as synthetic poliovirus. Yet, this did not seem to be the case for the vast majority of participants. As a result of the apparent lack of prior familiarity, before discussing the case of the experiment with IL-4 in mousepox we began asking how many participants had even heard of it. In any of the 23 seminars where this was asked, reported levels of awareness of more than 10 percent were extremely unusual.

This indicated state of knowledge combined with the pervasiveness of the theme of inevitability meant that many of the more detailed questions we wanted to pose about the content of professional codes of conduct or the responsibilities of scientists to prevent the deliberate spread of disease were poorly suited to generating discussion. To broach the question of what individual and collective responsibilities should be included in a dual-use code of conduct was too many steps ahead of participants' overall starting point.

As a result of the interactions in this first group of seminars, we made a number of revisions to the seminar designed to test out our emerging understanding of likely responses:

- In the spirit of publicly testing out thinking, we incorporated a new slide that indicated the themes from the initial five workshops. We wanted to present these to participants in later seminars to ask whether they concurred and then to prompt more questioning.
- From the start we planned to test out thinking regarding the appropriateness of seeking to stay ahead of security threats through rapid innovation. Initially we did this in a slide that listed (leaked) US biodefense activities that might well be interpreted as aiding offensive rather than defensive purposes. After using it, though, we felt that the discussion tended to be about a range of issues only loosely associated with evaluating 'running faster' and the slide presumed a great deal of knowledge regarding the appropriateness of such biodefense activities as set out by international agreements. So, a new slide was introduced detailing the recent massive expansion of biodefense programs in the United States (see page 27). This slide not only provided the basis for discussing the approach of 'running faster', but implicitly for questioning assumptions about the inevitability of scientific developments by bringing to the fore the matter of what gets funded.
- We had to cut the number of slides because of time considerations. Besides dropping the slide detailing the rapid advances in synthesizing

capabilities and recreation of the Spanish Flu virus, Slide 10 in Box 4.1 was withdrawn from the planned presentation and kept in reserve. This drew on analyses from the social studies of science to question whether science should be understood as consisting of the open exchange of ideas and resources. It was to serve as a means of probing expressed concerns that security constraints might jeopardize the open character of science. However, only once in this group were concerns raised about the possible undermining of the nature of scientific research cited. While participants suggested various negative practical effects from any restrictions, these did not refer to or allude to the character, ethos or norms of science.

Seminars 6–10

In this set of seminars, two of the audiences were significantly different in their composition from those in the first grouping. One entailed a departmental seminar that was open to undergraduates, which meant they significantly outnumbered senior researchers. The other was undertaken with a university center consisting of those trained in social and life sciences, but jointly concerned with the policy implications of genetics.

At a general level, both what was said and not said in this grouping of seminars were very similar to those in the first set. So there was little appeal to concerns about the possible erosion of the norms, culture, or ethos of science as such, few participants displayed prior awareness of the dual-use policy issues under consideration, many reasons were given against the viability of the pre-project oversight measures coming from the Fink Report, and attempts to determine scientists' individual and collective responsibilities through discussing codes of conduct proved rather unconstructive.

As previously, little support was given to the idea that the experiments in question or experiments in general should not be done because of security fears. As one participant said, even if results such as the synthesis of polio were predictable, 'until you actually physically do it, you don't know that you can do it. Like climbing a mountain, yeah, I'm sure we could climb up there because it's only a bit further than we went before, but until you actually do it you don't know that you can.' Again the importance of knowing and the inevitability of science featured as the most frequently voiced reasons against foregoing certain experiments. As one participant put it in relation to poliovirus: 'Surely the whole issue is the fact that this technology's been around for a long time and if Wimmer hadn't done it someone else would have done it.'

As mentioned above, in this group of seminars we included a slide detailing the multi-billion expansion of biodefense programs in the United States noted in Chapter 2. Picking up points about the importance of staying ahead of biothreats or general health threats through research, the slide asked whether we should always try to 'run faster'. We wondered whether bringing to the fore the contingent policy choices made about what gets funded in the life sciences (and thus what science gets done) would encourage some participants to query 'inevitability' claims. What happened was instead a general questioning of the US political rationale for their biodefense program and of the effectiveness of such a large-scale initiative (see Chapter 6).

As with previous seminars, the advisability of restricting publications was overwhelmingly doubted; prominent reasons for this included that publications could serve as 'a wake-up call', whether researchers published their findings or not 'the genie would have escaped out of the bottle anyway', and the large stock of existing knowledge already 'out there'.

In contrast, participants expressed a wide range of views about exactly *how* results should be communicated. Two main sets of reasons informed such contrasting assessments: one, the alternative evaluations made about what publications enabled by way of replicating research; and two, differing accounts regarding whether scientists today are compelled to actively publicize their research. The latter can raise key questions about the place of science in society. So when one participant commented on the 'need to have to hype up anything' in science, this was followed by:

Excerpt 4.2

MD: Um, why do we have to hype things?

P3: Well it's just that

P8: Called funding.

P3: Yeah, current competitive environment and just social situation we are in.

MD: So, if in the world in which you live funding is so crucial that if you have a finding which is interesting then you use it to the maximum in order that the funders know who you are.

P3: Yeah, the funding bodies want you to do that anyway, you know, depending on who's funding it. They want you to hype it up so that they get advertising for whatever, you know, their cause is.

MD: So even if you didn't want to hype it up, the funders might come along and say that's a very interesting piece of work that we've funded that you're doing, we'll put it into our big glossy magazine.

P3: Yeah, and likewise the magazines themselves want to sell. It's just this whole marketing environment that we live in nowadays.

P5: But, but

P9: I also think there's a social, I'm sorry I missed your earlier, I was unexpectedly delayed, but I think there's a social issue here, and if you've got their funding and you've used their funds for this sort of research, then you have an inherent responsibility to report to them how you've used their money. I mean the old traditional way was that if you, and you may well have covered this already with obviously a very excellent presentation, but the old adage was that it wasn't the responsibility of the scientist to govern what happened to the material, his obligation, or her obligation was to publish their findings. But it didn't move on to the area of responsibility to what then happened to that material. 'Cause you're talking a lot about negative things, but if you look for instance at dystrophin and this alerts one to the ideas of muscular dystrophy. It is very exciting for me to hear that people are being able to manipulate to this extent because the next thing is you would hope is that they would manipulate to benefit and, after all, I've lived long enough to see the outcomes of what's happened in the last 50 years and it takes time to turn to human advantage and there's no reason to think that the cycle, that that cycle isn't going to continue. That yes, you get negative outcomes, but then sort of following their trail you get positive outcomes. And are we in a position to make that decision?

(Seminar 8)

During the first part of this exchange, P3 contended that societal conditions and pressures compel scientists to engage in a hyping of their work. Herein, the contingencies surrounding whether research gets done mean regrettable actions must be undertaken. These contingencies about current policies in science contrast with the contention made elsewhere in the seminar that the production of results was inevitable and therefore there was little point in asking whether certain actions should be undertaken.

The last intervention by P9 suggested that upon receiving funding, scientists should be compelled to publish controversial research as there are possible benefits to be had from doing so. Upon subsequent probing about the implications of her claim she said, 'I recognize what ((P3)) is saying and I'm aware of the pressures, but I think you want to be very, if you're a scientist and involved in this you sort of, you want to be very even-handed with how you publish your work. You should publish all your findings up the same, you don't make a judgment on good for society or bad for society. It is the science that you are evaluating rather than future value to society or there's a future use to society.' To this suggestion another participant retorted:

Excerpt 4.3

> Well I disagree with ((P9)) there because I think it's a myth that I hope we've relegated to history the idea of value-free science. We've had a British Society for Social Responsibility of Science now for some 30 plus years and I think the sad thing is we cannot trust the politicians. And I understand in terms of, we also talked about the actual publications. To actually say there it is, you do with it what you want to. I know you're not quite saying that, but there it is, there are the facts and disassociate yourself and move on to something else is really irresponsible. I mean I, these three examples given here. What I'd like to have known before is the sort of ethics committees who looked into this work, I think you have to go back rather than deciding on what's happening afterwards. And the problem is the ethics is running a long way behind the actual scientific advances and I think scientists have to be at the forefront, they're not perfect, but sort of a Danish, well Scandinavian model with lots of people involved across the whole community, rather than leaving it to politicians.

Herein the significantly contrasting depictions were given for what science is like and what scientists should do.

As mentioned above, in the spirit of publicly testing out thinking, we devised a last slide for the seminar that indicated certain themes from the initial five workshops. Those included the aforementioned inevitability of science, the lack of expressed concerns about the potential demise of the character of science, and the seeming lack of knowledge of participants of dual-use issues. We wanted to present these back to participants in Seminars 6–10 to ask for comments, especially as many of the same points were raised in these discussions. The appropriateness of thinking about science as inevitable was queried through noting the disparity between the amounts of money going to tropical disease versus biodefense research. However, in each case we offered this sort of feedback on previous discussions, the responses were noticeably tense, limited, and in almost all responses rather peripheral to the points at hand. Because of these considerations we decided not to continue using this form of feedback and testing (see Chapter 6).

Seminars 11–17

The exchanges fostered throughout the third set of seminars were quite consistent with prior ones. In somewhat of a contrast with previous ones, concerns about the character of science being eroded through any policy responses were mentioned twice. In this set, the twelfth seminar was in a German university and acted as a so-called 'break group' to test for any radical differences in experiences.

Certain changes were introduced in the content of the seminar at the start of this set. To spend further time probing in discussions about what needs to be done in relation to publishing, publicizing, and choosing research projects, we cut out reference to the polio virus case. We instead asked the hypothetical question of whether the IL-4 mousepox experiment should have been done *if* the researchers had expected the results they obtained were likely. In order to move the discussion of the US biodefense slide away from considerations of the drivers of US science policy to testing out the logic of 'needing to know' and 'running faster' in biodefense funding, we included details about the chain of subsequent IL-4 pox work undertaken after the initial Australian publication (see page 52) as another manifestiation of the desire to know. This revision though did not shift the focus of discussion.

Much more successful – in the sense of generating lively discussion about the intended topic matter – were alterations made in questioning about the publication of research. Whereas in the past we asked whether experimental results such as those involving IL-4 mousepox should have been widely circulated, in this round we split publication into two parts. First, we asked whether the results should have been published in the *Journal of Virology*. This led to what were predictable, overwhelmingly affirmative declarations. We then asked whether the researchers should have communicated concerns about their results through the *New Scientist* as well. Here, in contrast, participants expressed a wide range of appraisals about how results should be communicated to non-specialized audiences. Given the importance of 'needing to know' regularly voiced by participants in relation to publishing in scientific journals, we were then able to inquire why this did not extend to more popular press outlets that were likely to reach a much larger audience. This was especially pertinent given that the vast majority of participants in the seminar had already indicated they were not aware of this experiment. This sequencing often produced quite lively and probing exchanges. So, to the matter of communicating findings in the *New Scientist*:

Excerpt 4.4

P2: What harm can it do? I mean, the public doesn't have access to the labs; the public couldn't do the experiment themselves. I mean who we should be worrying about are the people with the labs and the money, not Joe Public, surely.

P5: I disagree with that. Publicizing it in such a way, in *New Scientist*, the way that looks to me is the sort of title that someone latches on and creates hysteria with. And it's the sort of thing that's happened before with all the genetic modification in the headlines. Joe Public has had an effect in the way that that's being, research in that area is being carried out now, so potentially, doing it like that has an adverse effect on future research.

BR: So, would it follow from that, that you'd just avoid that sort of public publicizing in any respect, or

P5: ((Interjects)) No, don't avoid it, publicizing, completely. It's very responsible publicizing in journals where the people who want to see it can see it is fine but potentially opening it up to everyone who doesn't have the first idea what they're reading is a bad idea.

P2: I mean you could publicize it and explain it well.

P5: Yeah, but

P2: ((Interjects)) I mean, rather than not publicizing it at all, and allowing anyone to just go in and completely misunderstand it and explain it badly.

P6: I think as scientists we're told that we've got a responsibility in terms of public understanding of science, so we have to publicize as well as publish. If you like, we have to talk to the general public about the research that we do. Usually the Research Council grants – that's a requirement. But, you know, you have got to make sure that it's done in a responsible way. But, when you put something into something like *New Scientist*, you are in the hands, to some extent, of the journalist, who perhaps doesn't have the same scientific background, you

> know, sufficient scientific background, for a particular topic like this. So, you know, you've got to make sure that it isn't over-sensationalized.
>
> (Seminar 13)

In such free-flowing exchanges the competencies and responsibilities of scientists, science reporters, and 'Joe Public' were very much central to considering what should be done.

With the conflicting responses voiced about communication to different audiences, we wanted to ask if participants recognized that such topics involve highly dilemmatic choices and, if so, what can be done to take the situation forward. So, in this set, we devised a slide that listed the contrary responses offered as to whether scientists should widely communicate dual-use implications of their research. This was done in order to see (a) if participants agreed with the contention that this was a highly dilemmatic topic and (b) to then further collectively reflect on whether there were ways of dealing with this. However, again, this proved a conversation stopper.[29]

As part of thinking about how to move the conversation forward, we wanted to encourage discussion beyond the world 'as-is' by asking participants about the direction of future initiatives. This led to a revision of how codes of conduct and associated matters of individual and collective responsibility were introduced. In an effort to make the issues concrete for participants, we included a slide detailing the key questions identified in late 2004 by the British Ambassador to Geneva for consideration by national governments at the 2005 talks of the Biological and Toxin Weapons Convention.[30] As another, we asked whether participants agreed with the responsibilities set out as part of the World Medical Association's 2002 *Declaration of Washington on Biological Weapons*.[31] Yet, such attempts fostered little in the way of substantive discussion about community standards. In many respects this was perhaps not surprising given the overall relative lack of attention by participants to dual-use issues in the past and the difficulties of determining what should be done in any case.

Despite such limitations, by this group of seminars, we as presenters had become reasonably confident about likely responses and therefore knew how to sustain discussion in this type of group interaction. This understanding meant knowing how to continue conversations when they might otherwise be cut off. For instance, relations of expertise and status were potentially one way for the discussion to be closed down. Because of their prior knowledge of dual-use issues, specialist expertise, or position within a department, some participants could offer what would function as definitive accounts. Knowing how to prevent this from resulting in silence or nods of agreement was a key issue. This matter is explored more fully in Chapter 7.

Aside from particular slides that did not result in any significant dialogue, the content and structure of the seminars did facilitate wide-ranging debate for which we received many positive remarks. With this growing capability to generate fairly rigorous discussion about various dual-use issues, the question arose of how to proceed with experimentation. One radical step we debated was to begin by summarizing the conversation in previous seminars so as to launch into more in-depth policy discussions. This could also have helped test out our emerging assessment of likely responses. Undertaking such a major restructuring might have furthered the extent of questioning of the world 'as-is', but at the probable expense of reducing the overall quality of the seminars and closing down the time for participants to reflect on certain key issues.

Seminars 17–24

In this set, responses to questions about what should be done, published, or subject to oversight were in line with those reported above. There was, though, an overall greater expression that some sort of community decision-making framework should be in place to aid (but not regulate) individual researchers when they are considering difficult dual-use decisions.

The major exception to this overall consistent picture was a seminar conducted largely with first-year biology students, conducted at the request of a member of staff who organized one of our earlier seminars. Here the conversation was reserved, with the 'weight of gravity' of responses clearly tilted towards limiting publications, foregoing lines of research, and imposing oversight regulations. So one rather lively exchange following from questions about the advisability of the IL-4 mousepox publications went as follows:

Excerpt 4.5

P1: I think that's the wrong idea because you can debate why we should have the information.

P2: I think maybe, although they did make a mistake, um, it should have been published in case, for example, it got into the wrong hands it sort of looks like they are trying to hide something and you know maybe they should have, y'know, if they made a mistake they should be clear about it.

MD: So one argument here is that they shouldn't have done it and the other argument here is that if they didn't publish it then if it came out later they could be accused of covering it up? Yeah? OK.

P3: I think they should not have published it.

MD: Sorry, so if they, they should not have published it.

BR: Can you, because we want to just test out people's views about these issues could you elaborate a bit, like what would the reasoning be?

P3: There are insane people out there, they will use it.

MD: I, so there is another argument here saying they shouldn't have published that because people with malign intent could have gone and picked that up and thought about it and used it for hostile purposes?

P3: Yeah, yeah

MD: Anybody got a different view?

P4: I think that it should have been allowed to be published and all. It should have been like classified at least so a certain people should have had access to it really.

BR: OK so, so the argument here is that maybe it should have been just distributed just to a small number of people

P4: ((Interjects)) Exactly, I think it should have been like classified information like y'know only accessible to a certain group of people, you know if it, was, if they had to publish it I think that is what would have happened.

P5: So only on a need to know basis.

BR: That it should be published on a need to know basis, so as that's a, is that a qualification to what ((P3)) said previously about publishing it?

P5: See that is the question, you do biological research and you publish it for other biotechnologists, whatever, I mean, I mean I think it could be classified to other people that are not biotechnologists, there is a kind of a horrible group of people.

MD: Does anybody have a radically different point of view, somebody think of a completely different kind of argument?

BR: Well let me put the counter argument since the rest of our slides depend on it ...

(Seminar 23)

The stark contrast in our experiences with this audience of largely first-year students led us to reconsider the potential and limitations of the seminar design. While the content and the form of the seminars had been repeatedly modified in response to experience, this was done with a particular kind of audience: those with highly specialized scientific expertise, generally unaware of dual-use policy discussions, and often skeptical about the feasibility or necessity of proposed policy initiatives. By finding ways of testing out the thinking expressed by such a general type of audience, the seminars developed along a certain avenue. Efforts were made to unpack central notions such as the 'inevitability of science'. However, in doing so, to a degree we did not appreciate at the time, the seminar content became designed around challenging a limited profile of responses. When faced with an audience that expressed a different type of evaluation, many of the lines of questioning in the seminar did not have the same relevance. In this way we learnt about the limitations of what we had created.

On another matter, with the continuing difficulties in generating substantive discussion about scientists' positive responsibilities, in this final set of interviews we adopted a different approach. This was to provoke reflection on how responsibility was *distributed* rather than what it entailed. The content of seminars was overtly divided into two sections, one that considered dual-use cases and the other various possibilities for what might be done in response. The latter moved from large-scale institutional oversight changes to matters of individual action. As the first section of the seminar often elicited various responses to the effect that it was inappropriate to ask individual researchers to make complex decisions about what should or should not be done, the second section then asked where responsibility did rest. Widely expressed reservations about the feasibility or desirability of proposed policy initiatives then became the basis for inquiring back with participants about who should take responsibility. Thus the manner in which responsibility was constantly pushed about became a topic for discussion and an invitation for participants to reconsider earlier stated appraisals.

Discussion

In recognition of their group interactive feature, the results of standard focus group discussions are rarely quantified along the lines of surveys. Instead, recurring themes are elaborated. Consistent with such a practice, this chapter will not attempt to summarize 'what happened' through marshaling figures about how many times certain responses were given. It is possible though to identify broad themes of commonality which for pragmatic purpose can be treated as working generalizations.

As mentioned above, while changes were made in the content of the slides throughout the seminar process, we devised information and slides for all the seminars that broadly addressed three key issues in contemporary policy debates: Are there experiments or lines of research that should not be done? Is some research better left unpublished or otherwise restricted in dissemination? Are the envisioned systems of pre-project research oversight strategies sensible?

To the issue 'Are there experiments that should not be done?', the vast majority of responses supported undertaking the 'contentious' experiments cited as well as potential other ones. These provided knowledge and techniques that would be useful for a range of benign and defensive efforts. Yet, at a more basic level, the validity of asking such questions was doubted because of the said inevitability of science. Those that did query the advisability of undertaking certain research tended to be relatively young and most likely therefore to be students.

Likewise the advisability of restricting publications was overwhelmingly doubted; reasons for this included the importance of communication in countering the deliberate and natural spread of disease, the limitations of the details in articles to enable the replication of research, and the status of publications as just one way researchers share information.

Further, skepticism was expressed about the advisability of pre-project biosecurity oversight systems. While it was suggested that such a system might provide a necessary guard against outside interference in science, raise awareness of dual-use issues, and act as part of the needed reforms of wider university practices, the majority of responses were critical in nature. So such systems were deemed simply unworkable because of the impossibility of knowing the future implications of research, ineffective because terrorists would circumvent them, misplaced because British universities were not the types of places that should be causes of concern, and counterproductive because of the amount of existing regulations.

Becoming familiar with such responses about major areas of dual-use concerns, though, was just one part of the process of inquiry. In these seminars, we went beyond soliciting responses to particular questions to instead question how to question. This attempt was inspired by the work of Argyris outlined in the previous chapter. As acknowledged, achieving the forms of so-called Model II forms of learning he advocates is quite demanding and not something that can be realized through minding some simple technique. While being explicit about the goals of the seminars, and avoiding as best as possible unsubstantiated assertions, might help foster relatively robust forms of inquiry, such practices cannot substitute for sustained, trust-rich relations. The unfeasibility of achieving such interactions in our brief encounters meant we had to find ways of structuring the seminars 'off-line'[32] so as to promote our learning as moderators. A key part of this was to progressively and publicly question the predominant reasoning voiced. Initial and later formed presumptions by the presenters and participants were publicly scrutinized in aid of learning, a process which suggested various avenues for reframing discussions.

But just as many questions can be posed about the choices made in interactions, so too can questions be asked about how those are reported. The thematic account given in this chapter has been presented for the purpose of outlining a strategy of transformation within the seminars based on reflections of the interactions made at the time. Although relatively lengthy, this account has skimmed over many issues and interactions in favor of presenting an overall picture. Quotes from participants have been treated as more or less straightforward resources in making arguments.

By way of examining both substantive and methodological choices and challenges, the remainder of Part II considers in further detail a host of predicaments about how to conduct research and inquiry. This will apply to both life science research in relation to dual-use concerns as well as social inquiry into these issues. In doing so, the effort will be made to relate what is said with how things get said.

5
Openness and Constraint

Chapter 4 presented a broad overview of the central themes of the dual-use seminars. The analysis given formed the core of a number of academic and policy-orientated presentations to the National Academy of Science, the Royal Society, the United Nations, and others. Yet, it skimmed over many of the details and complexities of what happened in order to present a sense of the overall transformations in what was asked and what was said.

The remainder of Part II continues the previous orientation of considering contradictions and tensions in assessing what should be done.[1] Each of the following three chapters examine tensions related to both the substantive matter of the dual-use potential of life science research and interactional tensions associated with conducting inquiry into this topic. Those tensions include how to reconcile openness and constraint, expertise and equality as well as neutrality and bias. As will be elaborated, such a two-level examination is vital because of what was said depending on how things got said in the first place. So, unlike Chapter 4, this one will move away from treating transcribed excerpts as fairly straightforward analytical resources. Unlike Chapter 4, the next three are based both on analysis largely undertaken several months after the seminars were completed and on multiple readings of seminars transcripts and listening to audio recordings.

One starting point for this more in-depth treatment is the recognition that the recounting of general methods and overall trends is a poor guide to understanding situated interactions. So the previous chapter discussed how the seminars aligned and departed from typical methodological prescriptions for focus groups. Yet at this level of generality, many key matters for interaction were glossed over. While the literature on focus groups is often bereft of analytical attention to interactions between participants,[2] it is equally deprived of detailed analytical attention to moderators' moderations. This is a significant failing because moderator's actions are consequential and must be responsive to the particulars of each discussion. Even if we as presenters generally sought to promote dialogue rather than covert participants, choices were made in what was said. Harking back to the theme in Chapters 1 and 2 of what is taken for granted and what is taken as a topic of analysis, the choice about what to 'background' through interventions is as crucial as it is problematic. A goal of the following chapters is to redress the level of generality presented so far. The next section begins this by elaborating the evaluations made in the seminars and the reasoning underlying them regarding the themes of 'openness' and 'constraint'.

Openness and constraint in science

The stated desirability of openness

Chapter 4 concluded with a variety of generalizations about seminar responses. As contended, overall skepticism was expressed by participants about the wisdom of many proposed and current dual-use initiatives. Much of this doubt centered on the inadvisability of placing constraints on research. Certainly in relation to the initial slides and questions about publishing and conducting research (as opposed to the conversation generated from subsequent slides meant to probe these initial exchanges), little support was given to the idea of placing any security restrictions on research. Many of the concerns voiced were to the effect that impediments or limitations to what was done or communicated would jeopardize the peaceful and defensive benefits stemming from research. As mentioned in Chapter 4, in only one seminar (that consisting of largely first-year undergraduates) were the accounts given noticeably in favor of erring on the side of constraint.

To say the remarks in support of openness were in preponderance, though, is not to imply there was no debate about the undertaking and communication of research. So in one seminar, when it was asked in relation to the IL-4 mousepox case whether '... you think the Australians should then have gone ahead and published that result in the *Journal of Virology*?', this exchange followed:

Excerpt 5.1

P1: Well to state an opinion, I think research becomes useless if it's not published and that essentially is what we are all funded to do.

MD: So?

P1: So once it had been done whether it was intentional results or not it should be published.

MD: So they, this argument is that they have a responsibility, having done the work, received the funding for it.

P1: ((Interjects)) Yeah.

MD: Then the scientists' responsibility is to bring it to the public domain?

P1: Yes, because they have a responsibility to their own research team to have their effort recognized but also to their funding authorities and also to any other team that is working in a similar field.

MD: OK, any other thoughts?

P2: I think that's important that the responsibility to other people because if you don't publish the experiment then there is no reason why other people might not attempt to do exactly the same experiment, so you just don't go round in circles with people continuing to do the same experiment.

MD: So that's an argument for a scientific efficiency.

P3: I would think, while I understand what you are saying, I would think that it reduces scientists to automatons if that's effectively what they do and I would hope they should be willing not to publish if they didn't, if they really thought it was going to be a potential for bad things to happen.

MD: So they, then, this argument is that the scientist has got a responsibility over and above the science, the scientist has a responsibility to consider potential implications from publication.

P3: Definitely.

P4: I certainly don't go along with the argument that just because it's done we have a responsibility to your team to publish, that is a purely selfish responsibility, that we all have as scientists to knock up a number of publications that we get irrespective of whether it's good, useful, bad science or good science, this ((pause)) Our responsibility as scientists is to the community proper. If this had been done on smallpox, albeit if it had been done, OK I would guarantee that they could reproduce no ethical argument whatsoever for publishing that and telling the rest of the world we made a smallpox now which would kill all human beings. I mean that, we have to, we have to look outside to some kind of ethical context and that ethical context is not that we should publish just because it's been done.

MD: There was another hand up?

P1: Well, yeah, at the same time it must be a form of playing God though because people's ethics change so, so dramatically. I mean how do you judge what is ethical or not and it seems to me that if we are scared about a possible misuse that you have to have a really, really extreme case because you think about the people that would possibly misuse something. I mean they belong to a society, they belong to a group, it would be rare if someone who wanted to wipe out the entire human race, they would want to wipe out selected parts and clearly anyone that, that devised such a claim would also have to devise protection for their community and if they can do this then anyone else can come up with protection for their community. So I think, uh, you are just killing everything.

MD: So the essence of your argument is that there may be grounds on which you would think about not publishing

P1: ((Interjects)) It would be so extreme.

MD: Because it's so difficult to make that judgment it would be the extreme case.

P1: It would have to be something uncontrollable like if there was no antidote if you like yeah, yeah I mean.

(Seminar 18)

This exchange was one of the most mixed in relation to assessments of the advisability of publishing the IL-4 mousepox experiment. At stake in it were determinations of who scientists were ultimately responsible to and what that implied. Both P1 and P2 initially cited the said responsibilities of scientists to their colleagues, peers, and funders to support publishing in general. After P3 expressed concern that this reduced scientists to automatons, P4 contrasted the responsibilities to one's team and the self-interested motivations of individuals (which would favor going ahead with publication) with minding responsibilities to 'the community proper' (which might not). In response to these interventions, P1 modified her remarks to make an allowance for not publishing results for a possible, but unlikely, 'really, really extreme case'.

Yet, to the extent claims were made for supporting constraint of some kind in what gets published or done, they were overwhelming tentative and cautious. So, responses made in relation to the IL-4 mousepox case included:

Excerpt 5.2

P6: Yeah, I mean I think putting information out for people who possibly don't do their experiments in a controlled environment could make it, make it, the information, y'know, more widely available to misuse and things like that, and therefore you will get to people for wrong uses. So there has to be some sort of control on the way that it's um circulated, maybe specialists or maybe some kind of new set up so that certain people get sent the information.

(Seminar 9)

Excerpt 5.3

P2: I suppose it depends again on the detail. If it's a scientific paper with materials and methods which, I think it would probably, umm. If it was fully published and in a peer reviewed paper, it's possibly irresponsible to give full details. I don't want to, it's one of those, it's a real conundrum.

(Seminar 4)

Very few suggestions were made that specific experiments should not be done.[3] In only five seminars were any claims forwarded that the IL-4 mousepox[4] or the synthetic polio experiments should not have been conducted. The only other specific instances of no-go area that came up during the course of the seminars involved the examples of recombining flu viruses and certain work on 1918 Spanish Flu virus:

Excerpt 5.4

P4: Um well, I think there are quite clear ways that one could recombine parts of viral genomes in ways that wouldn't happen in nature, that you could predict would be dangerous, one obvious one that almost is happening in nature is recombining flu viruses to get the Avian Flu mixed up with the human flu and that sort of thing I think shouldn't happen basically.

MD: So you would have reservations about the work that was published back in August[5] where they recombined, they sort of combined bits of the 1918 Spanish Flu?

P4: Yes, I do, and I talked to a couple of colleagues at the time, I don't think that's a good idea, I don't think creating something like that in the lab ((pause)) basically I don't believe the biosafety protocols are strong enough to guarantee that sort of thing does not get out.

MD: So the reason you would be opposed to that being done would be on biosafety grounds?

P4: Yes. And on scientific grounds I see no reason to do it. ((Pause)) You've obviously got to balance the possible good from the experiments with some things that might happen in the future anyway, be prepared for it, I think that's justification then for creating something, if you want to know what properties it will have so that you may be able to defend yourself against it, that's the positive side. On the negative side is that you've created something that you know is dangerous and you're not so sure that you're going to be able to keep it when you made it, in that case I think it's more dangerous than what came out of it.

Even in this isolated example of a participant-identified possible limitation, though, the advisability of undertaking some of the initial work on the 1918 Spanish Flu virus was subject to multiple evaluations. These undercut the imperative of limitations on what was done. First, a concern was expressed about the strength of the particular biosafety provisions in place, then it was said there were no scientific grounds for the work, then that the negative risks outweighed the possible positive defensive gains.

A measure of the strength of the general preference for openness over constraint voiced in relation the initial seminar questioning can be indicated by noting that in all but one instance it was suggested (however tentatively) that certain things should not be published in the scientific press or done altogether, these were challenged by either those making the evaluations or by others present. So in one seminar, following two evaluations that the IL-4 case should have been published in the *Journal of Virology*, it was stated that:

Excerpt 5.5

P3: I don't know how it is possible but maybe there should be some kind of. Publish, publish part the data but maybe you don't say which sequences you say in your ((unclear)) that if you do something like that, that's what's happens, but you cannot do it in your kitchen because you do not really know. Publish it within the community so they can see what you did, but for the audience they do not really see what was done.

MD: If I understand you correctly then you publish enough so that the scientific community would understand what you said you'd done, but there would be important elements left out which would mean that somebody less scientifically capable would not be able to replicate?

P3: Or maybe for the scientific community itself also you can say you exchanged sequences but they do not say which gene so then you will only know what gene so you will only know that there is exchange somewhere which leads to this genome type.

P1: ((Interjects)) I think this leads to trouble because of an experiment, if the publication of an experiment is not reproducible it is not publishable. It is kind of a vicious circle.

BR: Just to pick up, but is the assumption that most research articles would allow you, if you read it, would allow you to replicate the research that's done, do you think that's fundamental?

P4: ((Talking at the same time)) fundamental criteria to publish, every article.

P5: ((Talking at the same time)) I think the same like ((P3)), because, so, if you want to do the same experiment you can um, you can do some co-operation with the group leader and get the information.

P4: I'm sorry, it's not about actually reproducing this experiment. It's just for the only guarantee that this experiment ever happened, there's no way of actually checking your data if you hand in your article and it gets reviewed. Publishing all your methods and everything in detail is the only, only guarantee that you really did what you published. I mean at least on the paper now you don't have any other guarantees. I mean this is really opening the chance for, for misuse of the whole publishing business if you limit your descriptions of the experiment and you can really publish anything. Today I did this, I don't actually tell you how, but I did it.

((Audience group laughter))

(Seminar 12)

In contrast to such consistent challenges to restrictions, in eight seminars I could find nothing approaching a reason offered for *not* publishing or undertaking the contentious work mentioned in our initial questions.

Openness and publications

Another measure of the strength of the general preference for openness is that the need for it was made in relation to contrasting evaluations about the ultimate importance of scientific publications. On the one hand, repeated assertions were made to the effect that once any elements of a research project become known, its full implications would be 'out'. So following his earlier claims about the inappropriateness of restricting publications, one participant argued that 'But you see, if you give the results, sorry if you give the results without the methodology then somebody can just go back and repeat those experiments, if they're heard of IL-4 insertion, you know they've got the technology to go and do that.' As such, the ready ease of moving back from reports of research curtailed the possible usefulness of any limitations. This conclusion is in line with many of the science policy analyses surveyed in Chapters 2 and 3 that spoke against the viability of modifying articles (e.g., by leaving out methods) because the results would be enough for others to replicate the research.

Yet, the general preference for openness was also evident in those responses assuming an opposing appraisal of the utility of publications:

Excerpt 5.6

P5: Just picking up on that, this is a discussion focusing on publishing as being some sort of, er, gateway to, er ((pause)) but I mean in science you know, people know these things are going on, people work in those labs. Those people tend to work there for two or three years and then move, so I don't think, I mean obviously publishing makes it open to the public much more, but in the scientific community if somebody really wanted to know about this, whether it's published or not, I think there are access to, into these sort of experiments. I'm not sure it's such a choice.

BR: So the logic behind that is, it's going to get out in any case whether it's published or not?

P5: Yeah, it's sort of arguing about the point of control is, because maybe the decision is whether it should be done or not. But I think once it's been done and methods are developed, it is not a secret if there are people doing it and these people will move to other places and take their expertise with them.

BR: So what follows from that is that a lot of controls that try to stop the knowledge, if you like, are going to be, ineffective, even if they were desirable, leaving that issue aside, they're probably going to be ineffective in practice.

P5: I think that the actual publishing once it's been done is beneficial because it opens the debate, so how you counteract it if it was going to be used.

P1: Well I was just going to say, you know, that the. I think it's quite relevant that these things are highly, highly specialized, and I think you can, even people within their own field would probably say that very often when you read a, an article that's

> published with materials and methods etc. and to actually deliver from an article which is a condensation of sort of five years, ten years work or whatever it might be, it actually can be practically quite difficult to do. And what you inevitably do is actually contact the other lab to try and get on board a technique or whatever. So I would sort of agree with the point that I don't think publishing is wrong necessarily and I think it's quite helpful. And also you have also to take into, bear in mind, I think, um the benefit of what that piece of research might tell you about other vaccines and response to viruses etc. which might be extremely useful for perhaps something like HIV or some other virus that might emerge that's unexpected. So you know I think it's a balance issue really.
>
> (Seminar 3)

In this example, P5 began by querying the idea that publications act as a necessary gateway for those within the scientific community. Scientific publications (such as the 2001 mousepox article) are necessary for *public* openness about science, but the movement of scientists means that experiments will be more widely known in the scientific community. Upon further questioning, he said that articles were a way of encouraging debate and action (presumably among the non-scientific public, politicians, etc.). P1 regarded publishing in general as beneficial and the mousepox article specifically as potentially extremely useful, but not because of any straightforward utility of articles *per se*. This was indicated by the initial preface that, in and of themselves, highly abbreviated scientific articles very often do not enable other experts to 'deliver' results. As such, the extent of disclosure achieved through scientific publishing would seem in doubt. What was concluded as following from this was that publishing is not 'wrong' in relation to concerns about the malign use of research. Yet, despite such remarks, P1 said that publications can be extremely useful in certain peaceful work.

Both participants' characterizations of scientific articles in the previous quote allowed them to be treated as important by alerting those outside the scientific community or by laying down a marker for follow-up by those within the community. Publications might even be helpful in furthering certain applications. However, articles were not terribly consequential for those with malign ambitions because such individuals could find out about research through other means or they would require further contact with the relevant researchers. Of note in this regard is that no one in any of the seminars applied the less than straightforward utility of publications to suggest that this curtailed their potential to aid the development of peaceful applications or to advance scientific knowledge in general. Further, at no time in any of the seminars did any participant make the following two-part argument:

1. Scientific articles have significant limitations in what information they provide;[6]
2. Therefore limiting scientific publications would not entail moving anyway from a state of 'openness' to 'limited openness', but rather amount to settling for a slightly different mix of disclosure and closure.

Such an argument, for instance, could have followed from the initial remarks by P5 in the last excerpt regarding the limitations of publications.

Instead of questioning the extent of openness afforded by standard practices, repeated contentions were made by participants that once an allowance was made for closing down something out of concerns about its application, the specter was raised of a wholesale closing down of science. So, for instance, that included concerns that:

Excerpt 5.7

P8: If you say you can't publish this then looking back we would never have dynamite and the other question, that I'm not sure but I think it pretends to this, is do we count scientific journals as, the press, in its wider sense because I would think 90 percent of the people maybe 100 percent would defend the freedom of the press. And yet we're now saying bringing in censorship and where does it stop? Do I say I don't agree with abortion and therefore you can't publish anything about abortion 'cos then because very quickly the atmosphere in scientific research and knowledge has thrived over the years suddenly becomes very claustrophobic and constrained.

(Seminar 22)

Blanket calls for openness had the significant effect of rendering immaterial vexed questions about how and who should mark the boundaries of where research of concern begins because they should not be drawn at all.[7]

The desirability of openness has its limits

As elaborated in Chapter 4, it was not the intent in undertaking these seminars to merely conduct a group survey of scientists. Rather one aim was to gain an initial understanding of the types of evaluations and reasonings offered by participants in order to subject the assumptions and inferences within them to progressive and joint scrutiny.

In relation to 'openness' or 'constraint' in the life sciences *vis-à-vis* communication, this was done principally through asking *who* should be made aware of the results of research (for instance, did this include 'the public'? as in Excerpt 4.4). As mentioned in the previous chapter, two additional approaches for communication were eventually included within the seminars to generate debate along these lines. The contrasting assessments offered of the appropriateness of these approaches facilitated moving beyond the binary of whether results should be 'published' or 'not published'.

In relation to restrictions about what research is conducted, the principal manner in which this questioning was done (more or less successfully) was through examining the importance of contingent funding decisions in shaping what research is undertaken. The tensions associated with what science should be conducted and who should know are addressed in Chapters 6 and 7 respectively when the relation between neutrality and bias as well as expertise and equality are considered in detail.

The certainty of openness (in the end)

While various suggestions were made regarding the preference for the lack of restrictions and self-determination in relation to initial questions about whether research should be published or conducted, as mentioned in Chapter 4, alongside such suggestions others were made contending that there was not much choice. In 19 of the 24 seminars, claims amounting to the 'inevitability of research' were voiced. Inevitability here meant either that any dual-use research of concern would be done (in the end) or that once done it would get 'out' (in the end):

Excerpt 5.8

P7: Once somebody has had an idea to do some research then the idea is kind of effectively out of the bag and it's very rare that there's only one group in the whole world whose working on a particular problem and you often get similar ideas popping up in several places all at the same time so, you know, if one particular group decides not to do it then there's going to be a group somewhere else who will do it somewhere so I don't think you should think you should pull back because of that.

(Seminar 20)

Excerpt 5.9

P2: I think in terms of the scientific community ((the choice made about how to publish one's research)) makes no difference whatsoever, because if you're working in this field a simple search is going to pull this out, the fact that you don't headline it doesn't make it any less accessible to the research community. Headlining it really only makes it more accessible to the more general media. Um, in terms of informing people who might be able to make use of the information, I don't think it makes any difference whatsoever, if you publish it you publish it and somebody who is interested in that area will find it, it would be very easy to find ...

(Seminar 7)

As suggested by these particular quotes, to regard something as unavoidable can justify quite definitive appraisals about what should be done. In parallel with the way in which the preference for openness could bypass vexed questions about the boundaries of research that was of serious concern, so too could the inevitability provide an all-purpose rationale that meant difficult issues could be sidelined. If the dual-use implications of research will become widely known once anything is published, then there is no need to entertain questions about exactly how one communicates that work. In its extreme formulation, 'inevitability' cut off certain lines of questioning. As mentioned in Chapter 4, the prevalence of this sort of response in the first set of seminars led us to relegate to irrelevance a slide probing about whether information on the *pace* of scientific developments in synthesizing microbes changed participants' evaluations.

Announcements of inevitability though were rarely as straightforward and extreme as to suggest there was absolutely no scope for decision-making of any kind. Rather what was contingent on choice and what was unavoidable were inter-mixed. Consider the following exchange:

Excerpt 5.10

P8: That's right, but picking up on ((P6's)) point as well, about the nuclear fission material work. I watched a couple of programs on Einstein on the atomic bomb a few weeks ago, and Einstein was beside himself that he'd decided to send a letter to the US president telling him that he thought consequences of the Nazis having the technology to build a bomb were such that he couldn't avoid telling the world about it. And, of course, they built a bomb and bombed Japan. But I think that was just ego, a certain arrogance there, because if he hadn't told the US president, he would have found out somehow at some stage, and I think you can't avoid these things happening. So just hoping to bury it only puts it off for a short period of time.

MD: So this is the argument that there's a certain inevitability about what happens in the scientific enterprise?

P8: Especially now as we have to publicize more about where it's getting all the funding and for the freedom of information etc., etc.

Herein while in his initial intervention, P8 queried whether Einstein's actions were really that consequential, this is qualified by noting that there was a time element to play for. Yet, when asked whether what he said could be taken as amounting to a claim of inevitability in science, the existence of this condition was affirmed by drawing on only recent changes to British science policies.[8]

The importance of the time-envelope generated by certain practices was a matter for debate. So in relation to a discussion about the case of British researchers that were said to downplay some of the implications of their research with IL-4:

Excerpt 5.11

P3: Just to go back to my point earlier, all you're doing with any of these methods like this one is that you are just buying time. In the end knowledge will out, I mean you can find the critical mass for plutonium on the web you see, that's not to say that during my lifetime restricting growth of nuclear weapons hasn't had some benefit but in the end all you're buying is time, I mean like any of these things, so it's the timescale on which you are interested in.

P5: Now I wouldn't downplay the idea of buying time because in a sense it is other scientific developments that find the chance to catch up, so let's suppose you make a killer virus by putting this gene in, if that's known to the scientific community but not more widely, um, in say the terrorist community, there is then a chance that a scientific solution can be found to this particular problem whereas if it comes out very rapidly that won't happen.

(Seminar 16)

Although claims about the inevitability of science figured in most seminars, just what was inevitable about science was conceived of in a selective manner. So, many acknowledged that new developments were ushering forth novel capabilities that might be able to facilitate the spread of disease. Yet, there was no suggestion that society had to get used to the certainty that current advanced science would be used with catastrophic consequences by terrorists, psychopaths, or states.

Openness and constraint in interaction

To carry on recounting exchanges in a matter of fact way is to treat them as analytical resources. As indicated in Chapter 2, this is just one way of orientating to such data. The account above could be supplemented with an analysis of the professional power of scientists to propose a purposely critical interpretation of the claims offered. The asymmetries noted in the arguments made would be prime starting points for such a reading, if they have not been treated in a judgmental manner already. Alternatively, instead of taking exchanges as resources, analytical attention could have been given to them as complex discursive phenomena. Following the lead elsewhere, for instance, the mix of 'empiricist' and 'contingent' repertoires employed might be examined so as to map how science is multiply portrayed.

As outlined in Chapter 3, a prime motivation in undertaking these seminars was to initiate a process of learning and experimentation regarding what should be done about dual-use research. That included questioning how to question as part of engaging in conversation. The focus here is not with representing 'scientists' views' or analyzing 'discourse' *per se*, but rather it is with exploring forms of intervention and interaction. The remainder of this chapter examines how the interactions in and between participants and moderators were organized with a view to understanding how that interaction entailed the negotiation of openness and constraint in the dialogue generated.

The organization of conversation

A key starting point for the analysis that follows is that conversations – such as those taking place in focus groups, in offices, or on street corners – are highly organized activities. Co-ordinating exchanges requires various interactional procedures. Among other things, parties to a conversation need to:

> organize the order of their participation – usually one person speaking at a time (turn-taking). They will fashion their contributions to be

> recognizable as some unit of participation – some 'turn construction unit' (turn organization). They will have practices for forming their talk so as to accomplish one or more recognizable actions (action formation). They will deploy resources for making the succession of contributions cohere somehow, either topically or by contributing to the realization of a trajectory of action or interaction (sequence organization). They will avail themselves of practices for dealing with problems in speaking, hearing and/or understanding the talk (the organization of repair). They will do so in a timely fashion (word/usage selection). They will do all of this with an eye to their co-participants (receipt design) and to the occasion and context, its normative parameters or boundaries of duration, appropriate activities and their order, etc.[9]

In relation to the seminars under scrutiny here, this co-ordination must be achieved for a topic that includes highly technical and contested considerations where two individuals 'moderate' interaction among an audience of scientific peers with varied levels of expertise.

Take the first requirement mentioned in the last quote, that of turn-taking. While initially perhaps a seemingly trivial point, in conversations there must be a way of determining who will speak next. How the problem of 'who goes next?' is settled varies tremendously though. In certain highly formal settings, such as debates or court room hearings, rules exists about who can say what and when. How turn-taking is managed has been a significant issue for those scholars examining talk across a range of formal institutional settings.[10] In traditional classrooms, for instance, that teachers play an essential (and asymmetrical) role in determining how and what turns are taken marks out the interactions in them from ordinary conversations. The typical dynamic of teachers being the questioners, students being the responders, and then teachers being the evaluators of those responses confers significant control to them over the dynamics of classroom discussions.[11] In brief, it is they who overwhelmingly 'select who speaks what, to whom, and when, but also [they can] hold or take back the floor for themselves whenever they wish'.[12] In fact, the unequal rights to speech within traditional classroom interactions has been said to warrant the interactional rule that 'only the teacher can speech in creative ways'.[13]

By contrast, in informal day-to-day discussions just who will speak next and what they will say is often negotiated moment-by-moment. Here, as opposed to institutional settings, more informal mechanisms are utilized to work out who will speak and what is speakable. In fast-paced, free-flowing conversations, seemingly inconsequential utterances such as 'um' or 'oh' can be vital devises in resolving who goes next. Also, in such informal settings it is often the case that 'topics can emerge freely, the participants are free to make diverse contributions to the subjects at hand, and anyone can initiate a new line of departure.'[14]

While distinguishing between institutional talk and informal ordinary conversations is a useful starting heuristic for thinking about how exchanges are organized, there are dangers in such a separation as well. One is assuming that the dynamics of all verbal exchanges in some institutional setting (e.g., a courtroom, a doctor's surgery) are necessarily of some particular institutional type. Discussions in such settings mix what might be labeled as 'ordinary' and 'institutional' types of conversational dynamics. Another danger is accounting for certain interactional dynamics by appealing to particular institutional considerations when they can also be accounted for by more generic, 'conversational' reasons. For instance, that doctors overwhelmingly ask more questions than patients might be accounted for through the high social status of doctors and the deference of patients. In other words, it is therefore a manifestation of doctors' control. Alternatively though, that disparity in questioning might be seen as the predictable outcome of one individual (the doctor) trying to obtain necessary information before providing assistance requested by another person (the patient). Seen in this way, there is nothing unique to medical consultations that accounts for a disproportionate questioning by one party.[15]

What kind of interactions characterized the seminars?

It is not immediately clear exactly what kind of conversational setting the seminars constituted. They certainly did not consist of throwing the discussion open to whatever participants thought germane. While the seminars were given as part of existing university faculty seminar series, they differed from the typical structure of these in important respects. Instead of following the usual lecture with audience questions at the end format, the 'presentation' was strung around a series of questions forwarded by the presenters. Unlike many seminars in undergraduate or pre-university classroom education, we were not testing them on their mastery of assigned work. With regard to many of the scientific issues discussed, the practicing researchers who made up the bulk of our audience could claim a far stronger basis for expertise and therefore challenge 'the teachers' in various ways.

The overall makeup of the seminars also shared similarities and dissimilarities with news interviews: they could be understood as a 'course of interaction',[16] wherein the dialogue consisted of sets of questions and answers that built upon what had come before and influenced what was said next. As we, the moderators, were directing questioning with a view to probing participants' thinking (rather than examining their mastery of some established knowledge), the seminars were more akin to interviews than classrooms. Yet, unlike most news interviewing, we as the questioners had a specialized knowledge about the policy topics discussed. As well, as *initiators* of the conversation, it was untenable to suggest we had no personal interest or knowledge about the topic.

The novel format of the seminars makes it difficult to regard them as akin to another type of institutional talk and therefore less straightforward to ask whether their interaction dynamics were due to institutional-specific or conversational-generic considerations.

Despite this rather unconventional format, the types of interactions were noticeably uniform. Some readers might presume that asking scientists about the potential for their work to further biological weapons might generate a range of reactions. Although tremendous variation might have taken place in the types of interactions fostered in the seminar (in relation to who spoke, for how long, and how they responded), this was not the case. This happened despite the fact that in contrast to some forms of focus groups, we as moderators did not try to stipulate ground rules for discussions at the start of the sessions. The relative predictability of substantive responses and the types of interaction fostered enabled us to plan how to revise our questioning strategy in order to progressively examine participants' reasoning.

The remainder of this chapter outlines the broad parameters of the relatively consistent interactions fostered with a particular focus on what kind of conversational turns were taken and by whom. This will be done with a view to considering how the organization of the seminars entailed a constant interactional negotiation of closure and openness.

Questioning and answering

One interactional characterization of the seminars is that they were rather civil and orderly events. As may well be surmised from Chapter 4, we as presenters very much set the terms of the conversation insomuch as we overwhelmingly chose the topics for discussion, elected who was to speak, and decided when to probe responses.

To elaborate, all the seminars followed a predictable pattern where we assumed the role of formal moderators and participants assumed the role of dutiful, and even generally deferential, responders to those questions. That meant only certain acts were done by us and them and deviations from this were often noted as deviations. So, no participant interjected to say when it was prudent to move on to the next slide[17] and only one proposed moving back to a previous slide.[18] Only once did a participant offer up an evaluation that the group's discussion had gone 'off track' and suggested getting back to the question posed. Only four times did a participant elect another participant to respond to the questions posed. Three times this was done with a prefacing acknowledgement of the expertise of particular individuals which accounted for why the selection took place.

The seminars were 'co-constructed' as a process of questioning by some and answering by others in that for each slide, participants waited for our initial remarks and question to finish before offering any response of their own.[19] While in ordinary conversation, pauses are taken as possible interactional points for others to speak, participants here waited for us to finish our remarks. Only once in all the seminars, for instance, did a participant interject in the course of our initial presenting of a slide. This happened while explaining initial proposals mooted at the time by the Medical Research Council, the Biotechnology and Biological Sciences Research Council, and the Wellcome Trust to develop checklists for grant committees, proposal referees, and applicants to identify potential dual-use issues. While proposing how that might work in practice, the exchange went:

Excerpt 5.12

BR: ... and that would have somehow (1.2) it is difficult to, to understand um which way that would sort of <u>cut</u> if you like, whether that would be a sort of bonus or, or a negative to research, but, but the, the idea anyway is that people will have to at um (.) least fill out a form asking them to think about these sort of issues (1.8) [and um]

P11: [It (.) it] smacks of lip service [to me.]

BR: [((laughter))] But I have not even finished.

AUD: [((group laughter))]

P11: [((laughter)) I, I chair our] our ethics committee and I know the seriousness with which people tick the ethics <u>box</u> and, and people don't sit down and weigh up the ethical (.) implications of their research, they quickly start ticking (out the boxes) and I sure this do exactly the same.

(Seminar 18)

Before considering the interjection by P11, it is important to comment on the transcription format of this exchange and how it differs from previous ones in the book. With the focus in this chapter turning to the details of interaction, it becomes important to not simply use transcriptions to provide the substantive gist of what was said. Rather, *how* exchanges unfold is also relevant. Just what sort of detail can and should be given of conversations for the purpose of understanding how they transpire remains a lively topic of discussion for those analyzing talk.[20] For the purpose of comprehensibility to those readers unlikely to be familiar with discourse analysis, I only use some of the most common notational conventions. In the transcription above stressed words are noted by underlining. Pauses are indicated with parentheses, with the estimated time of the pause indicated in seconds ('(.)' indicates a micro pause of less than 0.2 seconds). The words in single parentheses are the author's guess at what was said when the recording was unclear, whereas those in double parentheses are my description of what happened. Concurrent speak is designated by square brackets, with the start of anoverlap to the line above or below indicated by '[' and its end with ']'.

These additional details in the case of the exchange above are valuable because they suggest something of the transgression this interjection represented. So, P11 interjects with a condemning verdict during a pause before I could ask the audience the question on the slide. How P11 spoke (by interjecting before I finished) seemed to reinforce what she said and further indicated that her intervention was a condemnation. At the time this intervention threw me. Laughing was a way of coping with that surprise. That the laughter on my part comes quite quickly – before P11's sentence was completed and before the audience started laughing – is in line with my responding to the *timing* rather than the *content* of the intervention. The remark 'But I have not even finished' and the laughter that followed further indicates the transgression with convention in the seminars that the interjection represented.

The distribution of questions and answers

A sense of the seminars as largely a process of questioning (by some) and answering (by others) can be developed by considering in greater detail the number and types of things said. Table 5.1 lists the distribution of questions (Qu), formulations (For), and statements (St) given by the moderators and participants for a sample of half of the seminars. This listing excludes the introduction to the seminars and co-ordinating talk between the moderators regarding slide selection. That sample of 12 is

Table 5.1 Distribution of questions, formulations, and answers

	Moderators			Participants		
	Qu	For	St	Qu	For	St
Seminar Set 1						
Seminar 3	19	6	8	2	0	38
Seminar 4	27	11	23	11	0	81
Seminar 5	16	13	7	3	0	36
Seminar 6	20	10	18	8	0	44
Seminar 8	17	12	12	7	0	69
Seminar 10	37	8	23	15	0	45
Subtotal	**136**	**60**	**91**	**46**	**0**	**313**
Seminar Set 2						
Seminar 14	39	9	9	4	0	62
Seminar 16	24	3	4	3	0	40
Seminar 17	32	12	30	14	0	63
Seminar 20	35	9	14	11	0	61
Seminar 22	27	8	4	3	0	43
Seminar 24	24	14	7	4	0	60
Subtotal	**181**	**55**	**68**	**39**	**0**	**329**
TOTAL	**317**	**115**	**159**	**85**	**0**	**642**

in turn broken down into two groups of six taken from the first 12 and the second 12 of the 24 seminars.

'Formulation' refers to the act of 'summarizing, glossing, or developing the gist of an informant's earlier'[21] utterance. While these mechanisms are used across institutional settings by questioners for a host of reasons, the initial purpose in offering formulations here was to test for an understanding of what was said. They functioned more or less as questions because posing them implicitly asked individuals to accept or reject the (re-) representations made of their previous statements. This acceptance, rejection or refinement could then provide a basis for further comments or participants could pass on this opportunity. Excerpt 5.5 provides an example of a formulation when, after P3 spoke, the comment was made by MD that 'If I understand you correctly then you publish enough so that the scientific community would understand what you said you'd done, but there would be important elements left out which would mean that somebody less scientifically capable would not able to replicate?'

Some words of caution are necessary before considering the table in detail. Qualifying and classifying dialogue is tricky and necessarily subjective. Determining whether a given utterance is a question or a statement is not always straightforward. Remarks made in the form of a question can function as statements of fact and vice versa. For the purpose of Table 5.1 the utterances are classified according to how the author perceived their intended action in the exchange, whether that was to inquire or assert. Where there was doubt, the utterances where classified as questions. As well, assessing just what counts as an utterance is itself not easy. For instance, when individuals offered interjections during others' talk (such as a 'yes'), these could be taken as either meaningful statements of agreement or simple conversational acknowledgements of another's talk.

Table 5.1, then, seeks to provide a rough guide to what sorts of things were said by who. From this it seems reasonable enough to conclude that we as moderators functioned largely as questioners while participants overwhelmingly responded to what we said. So, despite the rather controversial nature of the topics discussed, the initial encouragement by the moderators for the participants to ask questions, and the scientific expertise of the audience, participants in the first set asked 0.15 questions for every statement and in the second this ratio was fairly constant at 0.12. Taking the formulations as type of questioning, during the first seminar set we asked 2.32 questions to every statement and by the second set that ratio increased to 3.47. My experience would suggest that this increasing ratio is down to our becoming more comfortable at probing individuals' reasoning, which in part had to do with becoming less defensive in response to participants' intervention. In other words, we felt less need to forward our own claims to keep discussions going and more comfortable at asking questions. That we became better questioners is also supported by figures that indicate while 43 percent of our statements during the first set of seminars were offered in response to participants' questions (many of which were for clarification, see below), by the second set that had increased to 55 percent.

To say something meaningful about the potential asymmetries in the conversation or how it blended openness and constraint in the kinds of things said requires going beyond bald figures regarding who asked questions and who gave answers. In general, it can be noted that our practice as moderators of being the questioners conferred an advantage in handling the discussion. This was because 'going first' in specifying what needs doing is (on the whole) more difficult than going later where it becomes possible to simply query the terms set out by a first speaker.[22] While in these seminars we were not trying to be argumentative in the sense of attempting to win around individuals to our own answers, that they largely responded to our questions did confer certain advantages. So we could easily offer a formulation of their statements which would then place them in the position of responding again.

This situation could have been much different. If, for instance, we had offered an evaluation about what needed doing as part of the initial discussion on each slide before anyone else had a chance to speak, then participants would have been able to pick at our evaluations during the subsequent discussion. This could have easily put the moderators in the position of the answerers and participants as questioners.

While there are certain interactional benefits in being the questioner, who asks the questions is not necessarily the same as who controls the way the discussion is subsequently framed. If, for instance, they or we offered strong challenges regarding the validity of each others' statements and the reasoning behind questions, this would have significant implications for the tenor and substance of the discussion irrespective of who was asking what of who.

While the extent to which we challenged participants could be subject to detailed analysis, at this stage it is perhaps more interesting to consider participants' challenges as they did not have a pre-established strategy for their responses as we as moderators did (though see Chapter 6). This, in addition to the likely contentious character of the issues discussed, could lead some readers to the conjecture that participants would have posed many types of questions. It further seems likely a significant number of which would have fundamentally challenged the framing of issues given by the moderators. Yet, rarely did this challenging take place. Bearing in mind the possibility for alternative classifications, of the 85 questions noted in Table 5.1 approximately 51 were requests for more information or clarification of points. Approximately 24 were questions posed directly to other participants or ones posed to the audience as a whole. Only about 10 were direct questions to the moderators that I would take as (in some sense) disputing or even testing what we were saying. These included, for instance, querying the artificialness of asking whether or not to publish the IL-4 experiment as well as asking where we intended to publish the results of our research given voiced concerns about the advisability of communicating about dual-use issues to a wider audience. One such instance of putting a challenge (of sorts) is given in the exchange in relation to publishing the IL-4 mousepox in the *Journal of Virology*:

Excerpt 5.13

MD: ... Should they then have gone ahead and published it (.7) is the question we're trying to get at.
P2: Yes.
AUD: ((Muted laughter))
MD: And the, the reason why they should have done that?
P2: The qu-question is why shouldn't they have done that ((unclear))? Why shouldn't they publish their result in the first place?
MD: So there, there is not really a question here that they, they had done the work, it was an unexpected result (.4) ah, they could see that there were implications (.) for possibly manipulating other pox viruses, but there was no reason why they shouldn't have gone ahead and done it, ahh and published it?
P2: I cannot see any reason at all.
MD: It is not even a question.
P2: Oh, not for me, personally no.

(Seminar 20)

Herein P2 does not just argue that the results should have been published (as many others did), but when asked to elaborate he did so in the form of questions back to the moderators which suggested little reason existed as to why any other course of action would be prudent. This questioning back then placed the onus on moderators to justify any other determination. Instead of doing so, MD responds with a formulation of P2's questions which then placed P2 in the position of having to answer a question. In this exchange, as in every exchange except one in the entire 24 seminars, an initial request by a participant for the moderators to respond to a question was not then followed by another such request. In other words, the moderators were almost never pressed to respond in some particular fashion. We could and did turn questions back to participants. This lack of being pressed about our reasoning is in line with the manner in which our position on any substantive discussions was basically a non-issue throughout the seminars (see Chapter 6). Noting such interactional points as has been done in this chapter provides a much more in-depth guide to the discussions than simply outlining the general prescriptive methods employed (as if so often done for focus groups).

Excerpt 5.13 is also of note for the manner in which it suggests what 'doing' reasoned university 'intellectual talk' is like.[23] While the IL-4 mousepox case slide eventually posed the question 'To publish or not to publish?'[24] and we asked whether the researchers should have published the results, on all but three occasions participants responded with more than a simple 'yes' or 'no'. Such extreme abrupt answers to questions in news interviews are generally associated with the questions being regarded as objectionable by interviewees and curt answers are often noticeable as out of the ordinary.[25] Here, too in the context of academic seminars, curt answers were infrequent and were attended to as out of the ordinary. In the case of Excerpt 5.13 and the other times where this 'yes' or 'no' response was given, the response was made noticeable by the audience laughter that followed. This laughter acted to lessen, if not wholly eliminate any suggestions of hostility on the part of the responder while still enabling the answer to have a definitive ring.

This section has sought to set out broad parameters regarding who took what sorts of conversational turns as well as describe something of how certain types of interventions were orientated to by those in the seminars. The analysis given has, however, only begun to elaborate the interactional dynamics in which dual-use issues were discussed. The next two chapters further this by considering how the key matters of neutrality and expertise factored into the types of conversations experienced.

6
Neutrality and Bias

The relation between science and politics is often fraught. In many policy and popular portrayals of science, it is frequently treated as the pursuit of objective truth. A certain method or methods of experimentation and validation are designed to ensure that individual personalities and prejudices do not affect what becomes accepted as authoritative knowledge. To the extent that politics finds its way into the research process, it is often depicted as a pollutant soiling an otherwise clean enterprise. Yet, perhaps especially in recent decades, the assumptions, agendas, and actions of scientists have received considerable public attention. Contention about topics such as global warming, animal experimentation, and the effects of tobacco have cast doubt on the advisability of regarding scientific claims as simply neutral, factual statements. When science policy comes under discussion – that is, how it is decided who gets public resources and for what – then it becomes quite difficult to maintain a strict divorce between the scientific and the political. Recent debates about the threats posed by bioweapons and the measures needed to combat them (Chapter 2) attest to how the stated agendas of scientists are intimately bound with political priorities of the day.

As maintained in previous chapters, social research and inquiry cannot be regarded as a neutral enterprise. Just what questions are addressed, how and by whom are contingent and consequential. Inquiry is a form of intervention in the world which is undertaken in some way rather than in innumerable others.

This chapter attends to the themes of neutrality and bias in research. As with the previous chapter, attention is given to both the substantive discussion regarding dual-use research as well as to the interactional dynamics associated with inquiring about it. The seminar dialogues are considered for the manner in which individuals embraced and distanced themselves (as well as the research process more generally) from attributions of what might fall under the umbrella terms of 'neutrality' or 'bias'.

As in the previous chapter as well, in this one the specifics of exchanges are treated as highly germane. A key starting point for the analysis that follows is the need to move beyond thinking of language as a simple neutral medium for representing the world. Instead, language is action oriented.[1] Exactly what is stated is contingent and consequential in the particular understanding it suggests.[2] A crucial question in examining the action-oriented aspect of interactions is 'why that now?';[3] that is, what is accomplished by a particular statement made at a specific point in time. Prior chapters hinted at a number of such action-orientated points about language. This chapter will extend such initial remarks by way of commenting on the play of neutrality and bias in the discussions.

Neutrality and bias in the conversations

The worlds of research evoked in the seminars often departed from a 'storybook image of science'[4] wherein scientists could simply be characterized as disinterested and impartial investigators pursuing truth. This is likely to be already apparent from many of the exchanges previously recounted. Excerpt 4.2, for instance, involved a discussion of the organizational pressures on scientists to hype up their research for the purpose of receiving funding, which was then followed by debate about the possibility/advisability of scientists assuming responsibility for how their work would be taken up. The said need to act so as to maintain funding was a recurring theme throughout the seminars, figuring in 17 seminars. It was evoked to make sense of both efforts to hype up and (much less frequently) maintain a low-key approach regarding any dual-use potential of research. In contrast, only once was a suggestion forwarded to the effect that excellent research would probably be funded whatever the wider funding context.

Related to this, the intent behind the actions introduced was likewise a recurring topic of discussion. As previous excerpts illustrate, reasons for scientists and policy-makers to have undertaken particular acts were sought by participants and attributions of intent were common in participants' reasoning. Just as any knowledge could be used for good or ill depending on one's motives, so too could actions be alternatively evaluated depending on motives. So a not untypical exchange went:

Excerpt 6.1

P1: What was (1.0) um, what was um (.3) Wimmer et al's intent in doing this, was it to ensure that it could be done and to cause a controversy ((inaudible)) or was it to um ah to (.5) do the science? Was it to do the science or (.3) was it um to stir up a hornets' nest?

MD: My =

P1: = How do you read it?

MD: My impression is that it was actually to do the science rather than to sti-stir up a hornets' nest. One has the impression that Wimmer was slightly surprised by the amount of (.5) hornets' nest that arose. Um (.) we we have in some of the conversations we have had with scientists um have people say, well, actually, it was necessary to do it, because although theoretically (.5) virologists knew it could be done until it actually was done ...

(Seminar 7)

The symbol '=' here indicates no intervening silence between the first statement by MD and P1's response. Such questioning (and answering) presumes intent to be a relatively straightforward matter that can be used to give a sense of meaning to action. Motives and actions were not always treated in such a straightforward manner, though, even when they were sought out. In a discussion of the merits of raising dual-use concerns to a non-specialist audience, such as the one that reads *New Scientist*, it was said:

Excerpt 6.2

P3: I um (.5) I disagree with that. Actually I disagree with that that route of (2.0) well there is (.4) you used the word pub-pub ah publicize but then again there is sensationalize and I think (.4) um if this was done before, if this wasn't picked up by the magazine from the journal, but was actually done before that to me (.4) perhaps illustrates that that, it was very contrived by the authors to perhaps (.2) have maximum impact possibly with a result of maybe getting more money for funding that sort of stuff. So I I think there was a conflict there (.5) of interest and I I personally would not have done that.

BR: So, so the negative evaluation stems from questions about the motives people would have for wanting to =
P3: [Tch]
BR: = initiate this sort of ah =
P3: = Yes because I mean scientists, of course, and actual researchers always want (.3) more funding and (.4) we're pressurized in many ways actually to, I wouldn't say sensationalize, but to ahh (1.2) 'sex up' is not (.) but you're, you're pressurized to make the most of the work that you've done (.3) um to gain maximum benefit from that. So I can understand why someone might do something like that, I just profoundly disagree with, because of the nature of the subject, I disagree with it.

(Seminar 22)

Here the initial and follow-up response by P3 indicates that while he profoundly disagrees with non-specialist publications coming out before the related scientific article, he recognizes that researchers are pressurized to do things that might otherwise be judged negatively. His search and the abandonment of the search for a phrase to label what researchers are pressurized to do suggests a basic difficultly in characterizing certain ways of behaving.

As mentioned in the previous chapter, though, 'inevitability' was often said to rule out choice and therefore the relevancy of intent. In relation to matters of neutrality, this meant certain questions were deemed 'academic' and thus not important issues for reflection. Regarding the advisability of communicating dual-use research beyond scientific audiences, for instance, one participant echoed comments elsewhere in saying 'I think once you publish in an academic source then ((more public attention)) can follow whether you want it or not. I mean whether you promote it or not is a side issue really.' In 13 seminars this was said to be the case because the knowledge would eventually be out due to the trawling of the scientific literature by magazines and newspapers. It is worthwhile to comment, however, that such sentiments were taken to justify alternative paths. One was to treat the extent of public promotion as a relatively unimportant side issue as indicated in the previous quote. Another was that 'There would be absolutely no point in trying to keep it within that one research community, it would not have been possible, so to circulate it as widely as possible I would have thought was by far the best strategy.'

As noted in the previous two chapters, the extent to which the said unavoidability of certain research factored into discussions and the manner in which its evocation brought them to a close led us, as the moderators, to seek ways of testing out individuals' underlying presumptions. Chapter 4 described how by the end of the first set of seminars, one way we sought to do this was through introducing a slide specifying the multi-billion expansion of biodefense programs in the United States post-9/11. Such a slide offered evidence that it was inappropriate to regard what research gets undertaken as simply inevitable because significant choices are made in what gets funded. Yet instead of this slide bringing this issue to the fore, in 15 out of the 20 seminars in which this slide was used the peculiar politics of science funding in the United States figured. Less polite exchanges included:

Excerpt 6.3

P2: They actually don't mean biodefense do they?
MD: Wha
P2: They mean bioattack (0.3) actually.
P3: After all this is George Bush
((Laughter and inaudible talk))
P2: We we are going to develop it before you bastards do. No I think that's outrageous. My <u>per</u>sonal view is that that kind of <u>excessive</u> ah attention to this area is is is a public outrage, ah I mean if it happened in this country, on a small scale because it would never happen on <u>that</u> scale in this country, I I would anticipate ah (1.2) public concern about a disproportionate amount of money was being spent.
MD: But it's disproportionate in that it is sucking scientists
P2: Away from, ah, much more appropriate research. Ah it's based on scare mongering and and working to a very political agenda (.4) and I would be deeply suspicious of it.
MD: It is [a]
P2: [Bu]t I would would get a piece of it as well.
((Group laughter))

(Seminar 7)

in addition to:

Excerpt 6.4

P3: ... It just smacks to me of being politicized you know I mean that is a political, that is <u>not</u> really a scientific decision, that is a political decision to make that funding and I think (.5) ahh it's very difficult, I mean it's great for the scientists that are working in the labs I mean, foo they probably have ((inaudible)), but it's a political decision isn't it? I mean they are funding purely for political reasons based on paranoia or whatever you want to call it, you know security, national defense, and (1.0) that might not get the right science but it's probably going to get (.) someone re-elected by saying we spent this much on making sure

BR: So, so, so I hear you saying a number of points that have been expressed here as well that yes =

P3: = I guess really what I'm pointing out is the political dimension it seems to me, for that amount of money, and and I-I think we need to differentiate (.4) what scientists do because of scientific reasons and what politicians do for political reasons and that mangling process where the two collide because scientists want money and they want funding and they want to do research. Politicians want to get re-elected. Quite often you end up with a horrible monster.

(Seminar 22)

Excerpts 6.3 and 6.4 indicate a rather complex picture of the politics of science funding, wherein researchers regard themselves as complicit in questionable actions.

Neutrality and bias in analysis

What might be made of the highly 'interested' dimensions of science noted so far? In general terms that question has been commented on in various ways throughout the book. So, it would be possible to use such accounts as straightforward analytical resources to make arguments about the political dimensions of science. Herein, an analyst would want to assemble as many quotes as possible that lent weight to a certain characterization of the relation between science and politics. Alternatively, it would also be possible to examine the discursive strategies whereby participants move between accounts about the 'politics' and 'non-politics' of science. Herein, an analyst would want to assemble together as much data of an individual's talk as possible to develop a sense of the varied ways these were related.

As discussed in previous chapters, though, these seminars were undertaken with the practical concern of how to promote inquiry. As such, what was sought within the seminars was to identify likely matters on which participants would have different appraisals and then question these in order to draw out in public the underlying assumptions and inferences. As the slide about US biodefense funding was placed towards the end the seminar, we were able to contrast statements made about it with responses to previous slides. For instance, in 12 of the seminars participants expressed concerns that the knowledge generated through the US biodefense program would have a detrimental affect on security because it would generate public knowledge on the very pathogens that were likely to be used in bioattacks. In six of these the suspicion that the anthrax letters of 2001 originated from a source within a US government establishment was used to support this argument. This likely origin meant that it is not just the actions of outsiders to the biodefense program that should be of concern. Yet such assessments of the dangers of biodefense-generated knowledge seemed to diverge from earlier claims in the seminars about the importance of research in staying ahead of threats (e.g., the work with IL-4 and mousepox as well as synthetic polio virus). This potential divergence could then become a topic for discussion:

Excerpt 6.5

P5: I mean I find it extremely <u>frightening</u> <u>really</u> because (.) they're making all of this information, you know. They are the ones who are going to be generating this information and there's nothing to ensure it's not going to get out and the anthrax scare in in American came from where? From within America, from within a government institution.

MD: This is a (.) different kind of argument. This argument is (.) one which is saying that they are actually in<u>creasing</u> the possible threat to themselves =

P5: [Indeed]

MD: = by pouring all this money into this particular area. Because all of it's going into precisely the agents that (.5) you would be concerned about

P1: That's right.

BR: OK. But is the thinking there (.) because we've had up earlier slides and people said sort of y'know that knowledge is a good thing kind of argument. You need to know. And now here there's some sense that maybe you don't need to know so much. So could pe-people just say, try to y'know elaborate the assumptions I mean. What is it that =

P6: = What's going to happen is these labs are going to make groups of people that are answerable to no one apart from themselves

(.7) and the people that are paying them the money (.3) and no one from the outside world will know what's going on.

P7: It is the difference between knowledge and propaganda.

P2: ((Inaudible))

MD: Well remember most of this is NIH and most of it will get published. Because ((inaudible interjection)) as you said people are not going to be prepared to spend years on research programs if they don't publish it, could you?

(2.5)

BR: Is the threat, sorry to go on about this, so, I mean, so this person who said it was propaganda, it's the fear let us come back to that. So is it, it is not the knowledge. Is it that the knowledge is dangerous? Is that what people are thinking? Or is it just dangerous knowledge because it's just feeding at the frenzy of fear? What? Could people =

P7: = You can't divorce the two by putting something between it. People will only (.3) the scientists will only work on things which we're willing to pay them to <u>work</u> on. If they're they're not being paid to work on these things, they'll go and work on something else which they are paid to work on. If we paid them to sort of find a decent cure for sleeping sickness which kills a damn sight more people than AIDS in some parts of Africa =

P2: Mala[ria]

P7: = [th]en um they would actually be doing some good. However, if they're not being paid to do things like that, therefore they'll not work on it.

(Seminar 4)

With the often repeated statements about the importance of doing and communicating research combined with critical remarks about the United States taking up a substantial biodefense research program, interventions were made in 10 of the 20 seminars where the biodefense slide was used that researchers could redirect the biodefense funding toward more 'appropriate' ends. So, following comments regarding the concerns about the misallocation of resources given that 'people are dying of AIDS and cancer', one participant remarked:

Excerpt 6.6

P1: Yes, I mean, except that it depends. The actual research which is going on is not that focused actually (.) um. Inevitably if you're going to try and spend that kind of money. I mean I know people that have got grants under this program, I know the kind of work that it's funding. Um inevitably if you're going to ramp up your spending at that rate you're actually going to have to bring in an enormous amount of of (.5) interest that is already existing in the community (.3) in the general area of infection and ahh you know, pathogenesis of infectious disease, immunity to infectious disease, all of those. People, y'know everybody who worked in that area suddenly sat down to their desk and thought how can I get a piece of this money. Right.

((Group laughter))

P1: That's exactly what happened.

BR: Sure, sure, but is that to say it's not necessarily a bad thing, I mean what's the implication of your argument?

P1: Well my implication is you could just as easily classify, I would guess at least 80 percent of this research, y'know as being nothing to do with bioterrorism. Umm.

BR: So not not a worry about diversion that people have.

P1: I mean yes exactly, I'm not particularly concerned about that aspect. When when we had the AIDS program a few years ago exactly the same thing happened. It's not possible for government to create a new research agenda in a period of, of that size, in a period of two years, it simply can't be done. All that happens is that people think of new ways of repackaging stuff that they're they're already interested in doing. Ah that's largely what goes on. May- there is some change, I am not saying there is not, there is some diversion, so instead of working on, y'know let's say um (1.2) immunity to, I know somebody who was working on immunity to um (.8) um (1.0) some gut pathogens. I forgotten which one they were working on before. And they switched pathogen to working on one that they knew would be funded under the bioterrorism program. But what they were really interested in was was gut immunity.

(Seminar 14)

So against concerns made about the diversion of resources away from diseases such as AIDS and cancer, P1 contended that the agendas of science were not as directable as might be counted on in funding programs. As in the case of AIDS funding in the United Kingdom, what largely happens with step-wise increases in funding is that scientists carry on with their previous research. This suggestion somewhat contrasts (at least in the near term) with the final statement in Excerpt 6.5. In any case, the exchange in Excerpt 6.6 was followed by an acknowledgement by a graduate student of the possibility that the funding could serve ends other than the stated one, but with a questioning of the ultimate wisdom of such a course of action:

Excerpt 6.7

BR: Does anyone else want to offer a view, um (2.0) of this? I mean, a number of people said before that we need to know, that knowledge needs to be out there, maybe someone's come in with a different suggestion in the back here,[5] does anyone want to offer up any evaluations of this?

P10: Just like ((P1)) saying about 80 percent of the research is really (1.2) semi-relevant to the topic, it's more a matter of presentation of the work. I mean, as it's written up there, y'know, it seems like it's mirroring, pretty much mirroring the development of the Cold War style arms arms race. The difference is it's against ah a a completely unknown (.5) quantity. We're not really sure if or what the enemy is or what they're doing. So in that sense it's kind of creating hysteria. But I mean (1.8) I'd say that if it's presented to the public under this, everything we're doing is all in the cause of biodefense, what kind of perception does that give the public of science ah in the longer term? I mean we've got a massive hangover from the Cold War as scientists, people are very cagey about everything they, y'know you read about there is a general mistrust of science in in the United Kingdom, I think =

BR: [So, so]

P10: = [and] and I don't know if this would have a good or a bad impact. If every time they hear about what biology's going on and it's always to do with bioterrorism (1.0) rather than saying well actually this is about these gut pathogens that could help people in other countries, it is not really going to [form the right connections.]

BR: [So so your your] So your worry is if if science becomesassociated with issues about weapons or these military issues or [anything]

P10: [Well] I think this is a very handy way, like ((P1)) said,it is a very handy way of gaining extra money as a scientist. So it furthers your immediate cause but I question whether the long-term effect on the subject (.5) is such a good thing.

> But I don't think it is a matter that any individual scientist has power (.4) to um affect.
>
> (Seminar 14)

In such exchanges, the reasoning for the alternative appraisals of programs of funding become matters for reflection and further deliberation.

While it was possible to promote dialogue through the use of a specific sequence of slides meant to probe participants' expected responses, what was less successful were attempts to directly challenge responses. As mentioned in Chapter 4, in Seminars 6–10 we prepared a final slide that included reflections on what we had heard in Seminars 1–5. In this we disputed the widely expressed notion that scientific development should be regarded as inevitable by first pointing out the disproportionate funding for biodefense versus tropical diseases, and then second, by asking what interests should shape the agenda for science funding. In addition, we asked whether the seminars we conducted could be regarded as 'part of the problem or solution?'; this in the sense of whether they were fueling inappropriate preoccupations or engaging in necessary dialogue.

However, presenting such points did not generate much discussion, in fact it tended to stop whatever dialogue had been fostered up to that point. In one seminar nothing at all was said in response. In two seminars a number of interventions were made, but had only three statements between them which the author would regard as relevant to the matters proposed. In another a point of clarification was asked and concerns about the over-regulation of science expressed. In the fifth, funding priorities become a topic of significant discussion, but not in a manner that directly spoke to the theme of inevitability. Beyond this general lack of responses, the awkward silences and tenor of statements lead us as moderators to drop this approach for testing out participants' responses.

Questions about questioning

Mentioning the adoption and then withdrawal of such a feedback slide speaks to the range of possibilities in how the seminars could have been conducted. In relation to the themes of this chapter, overtly challenging statements made by participants most likely had implications for their assessments of our predispositions and impartiality. Such a line of pursuit certainly differs from typical prescriptions for how focus groups should be conducted. In these, moderators are meant to be non-judgmental so as to be seen as neutral. Yet, such blatant attempts to evaluate responses as represented by the withdrawn feedback slide are just one of the most obvious ways in which the personal positions and predispositions of the moderators were potentially at stake.

Exactly what we said could have been taken as indicating some kind of bias. As elaborated in the Introduction regarding news reports of research into the 1918 Spanish Flu, what is put in and what is left out of descriptions can be highly consequential in the understanding fostered. As such, what was included and excluded by us could have led to queries or criticisms regarding why we did so. Just as analysts studying talk might ask 'why that now?',[6] participants in seminars might ask the same. For instance, in talking about the US funding of biodefense, any number of points could have been, but were not, included by way of background. While we focused on NIH-sponsored initiatives because of the magnitude of funding involved, their relevance to academic researchers, and the transparency of the funding programs, it would have been possible instead to focus on defense agency related initiatives. Likewise, we could have directly commented on whether NIH biodefense research was being bought at the expense of other civilian research. Just how relevant these or other considerations are to a discussion about dual-use research is a matter for disagreement.[7]

If queries could have been raised about the background information we supplied, so too could they have been raised about the questions we posed. As elaborated in previous chapters, a number of binary choices were initially use to promote dialogue (e.g., should you publish or not publish experiment x?). That these questions offered a highly simplified framing could have been noted and the reasons for this pursued.

In addition, there were continuing tensions about if and how to keep the conversation going. In the absence of contrasting participant responses to proposed questions or because of participants' requests, for instance, we as moderators could have stated counter-arguments to what was said (see, for instance, Excerpt 4.5). Yet, since these counter-arguments would overwhelmingly be made in opposition to participants' interventions that stressed the 'need to know', repeatedly stating them might make us seem to favor a certain line of action. Likewise, the extensive use of formulations by the moderators to summarize and develop the gist responses could have been queried regarding the gist we forwarded. Even small details, such as the strength, speed, and tone of remarks of 'yes' or 'no', might be taken as indicating preferences. Of course, reasons can be sought for why anything at all is being asked.[8]

In short, our contributions could have taken a variety of forms even if we as moderators were otherwise faultless in conducting ourselves in a manner that promoted open discussion. Yet we as moderators were not faultless in this regard, as the excerpts included in this book attest.

In the absence of queries

Against this set of considerations, it is noteworthy how rarely participants brought attention to our possible agendas, preoccupations, or predispositions. Consider several points in this regard. First, over the course of all the seminars no one raised questions of any kind during the introductory phase when consent was asked for recording. Second, in only four seminars were we asked about our own positions on the specific questions posed or for further details about what we aimed to achieve through our work beyond the points we made when introducing the seminar. As far as possible to tell, two of these requests were made by undergraduate students.

Third, in general it is quite difficult to find occasions where participants' *statements* overtly made our possible agenda or positions topics for discussion. Perhaps in up to eight interventions this was done in a more or less subtle fashion. One of the two exceptional explicit and direct allegations was made in relation to a question about the advisability of synthesizing agents where one participant said: '... I mean look at the flu, for example, I mean that changes every, some flus change virtually every month. Shouldn't we trace flu round using this sort of technology? So you're damning the technology just because it happens to be able to make Ebola potentially in about three or four years' time.'

Fourth, in about ten other responses critical remarks were made about the questions being asked or the background information presented that suggested a reframing of the issues was required. Such comments *might* have been meant to imply the moderators were pursuing issues in a highly contingent manner because of our particular concerns. So in relation to asking whether the IL-4 mousepox experiment should have been done if the researchers had regarded the results they obtained as a likely outcome from the start, it was commented that:

Excerpt 6.8

P1: I'd, I'd like to throw the whole thing open by saying (3.0) it would be (1.0) it feels very artificial to pose this question in relation to <u>that</u> (.) this particular project, but maybe this is a sort of question we should be finding social and political mechanisms for asking about a <u>lot</u> more that goes on in science.

BR: What's the, where's the artificialness come in in terms of?

P1: Because, because generally we don't ask the question (.) should this have been done (3.5) and there's no (.4), as it were, more public mechanism or modes for doing that. So so to do it in relation to this seems to me very artificial, but to indicate that there needs to be a way in which (.) ah (.) broader, there could be broader input and <u>reflection</u> about what is done in science seems to me very important and it highlights the absence of those mechanisms.

BR: OK, so wrong question. Any anyone else want to give an evaluation or even talk about these sort of (.6) underlying issues that might be um relevant?

(Seminar 6)

The critical remarks made by P1 suggest the attention to individual decision-making provides a highly artificial basis for thinking about the issues at stake. However, this does not appear to lead to any straightforward allegation against our reasoning for why we as moderators did this. Where doubt was cast on the soundness of the questions or background information, our responses included: letting the next participant speak, acknowledging certain faults, asking about the implications of the participants' comments, offering conversational acknowledgements that then put the next response back to the participant, giving justifications for the course pursued, or asking about the reasoning informing participants' comments. What we never did, though, was suggest the particular questions asked or information presented derived from biases in our ways of thinking.

It is also noteworthy how rarely participants made each others' possible preoccupations or predispositions topics of conversation. Although the excerpts presented so far make it clear that participants did not always share the same evaluation of what should be done or a sense of the principal issues at stake, at no stage did any of them seek to explain the response of another through citing some underlying motivational factor or political predisposition. While it seems quite reasonable to expect that at least some participants could have commented to another that 'You would say that wouldn't you, because of' some personal consideration, none did so. Such an ascription, of course, may well have been taken as a hostile personal attack. Yet, while there was sometimes passionate debate, this did not result in allegations of other participants' bias.

Taken together then, the sections of this chapter suggest a marked contrast: while concerns about the motives and wider politics behind those involved with dual-use related initiatives were commonly expressed by participants, this did not apply to others within the room. This lack of scrutiny was in line with the lack of questioning of us in general as detailed in Chapter 5.[9]

While it is possible to attribute the lack of questioning about agendas and predispositions to individual preferences to refrain from making awkward or embarrassing points to other people present, it can also be seen as a likely outcome of the type of interactions we engaged in. In a study of the interactive dynamics of news interviews, for instance, Clayman and Heritage asked the question, 'how do interviewers manage to assert themselves in an adversarial manner while maintaining a formally impartial or neutralist posture?'[10] While journalists are generally expected to be adversarial and impartial, combining both in practice is problematic. Since what is asked could be done in numerous other ways, just what interviewers say on any given occasion is open to criticism for how it departs from some idealized standards. Yet, what is remarkable is that despite the potential for accusations of bias, news interviewees rarely make such overt personal accusations.

Clayman and Heritage identify two facets of interviewing that are central in both minimizing accusations of bias and providing a defense against them when they are made. One is the basic question and answer format of interviews. So long as interviewers formulate their interventions as questions, it becomes quite difficult to substantiate accusations of bias. In response to such charges, interviewers can claim they are only fulfilling their professional role in asking questions. As Clayman and Heritage argue, because of this readily available get-out mechanism, interviewees rarely make such claims.

A second facet is the practice of speaking on behalf of third parties. So, as mentioned in the Introduction, a report about the work on the 1918 Spanish Flu in *Nature* stated that while some regarded that work as highly beneficial, '... others have raised concerns that the dangers of resurrecting the virus are just too great. One biosecurity expert told *Nature* that the risk that the recreated strain might escape is so high, it is almost a certainty.'[11] Such a reporting of varied positions is central to 'balanced' news reporting. That the claims noted are said to be made by others is vital in distancing journalists from accusations that their own preoccupations and concerns are entering into the analysis. In news *interviews*, however, evoking third parties has an added significance. In the dynamic back and forth of unscripted dialogue, being able to attribute particular ways of thinking to third parties is vital for interviewers to be able to justify their contingent choices in how to question and to be able to distance themselves from any adversarial overtones associated with those choices. Even anonymous entities evoked through phrases such as 'critics have said X' or 'the public has expressed a concern about Z' often enable interviewers to effectively side-step accusations of bias while challenging interviewees' statements.

The need for news interviewers to reconcile the tension between being neutral and questioning parallels the concerns identified in Chapter 4 regarding how moderators reconcile tensions in the course of focus group interactions. As professional questioners, both must make choices in how to question, the possibilities for which constantly arise during emergent dialogues.

In general terms, we employed a similar type of question and answer format and the practice of speaking on behalf of third parties. As elaborated in Chapter 5, in the seminars the moderators were largely the ones asking questions and participants the ones giving answers. In addition, the process of constantly revising the seminars provided a way of building into the dialogue a probing of the predominate responses offered while reducing the need to resort to impromptu questions that might appear motivated as a challenge to participants' claims. It is worth noting, though, that the revisions were undertaken in part to find those questions and issues that would generate varied evaluations between participants.

Our dual-use seminars also entailed quite regularly speaking on behalf of third parties (see, for instance, Excerpt 6.1). This was the case because the scripted contribution largely consisted of offering accounts of activities happening 'out there' in the wider world. So, for instance, a not untypical background statement regarding the IL-4 mousepox experiment went as follows:

> what it did in fact was to close down the cell-mediated arm of the immune system and they ended up with a recombinant virus which killed mice genetically resistant to mousepox even when they were immunized against the mousepox. And it didn't take very long for people to say (.3) hang on a minute, what if somebody was to do this with smallpox. So that was the basic mousepox experiment and that leads us then to our first question ...

To have said that 'people' (or for that matter the 'security community', or 'policy makers', etc.) began to voice concerns performed a number of functions: it underscored the importance of the issues mentioned, it helped shore up the legitimacy of our asking certain questions, and it allowed us as moderators to raise concerns without closely associating ourselves with them. Clearly though, alternative ways were possible to present the same issues. Instead of attributing concerns to others, we could have sought to make our personal assessments central to the accounts given. So, the background to the IL-4 mousepox experiment could have been:

> what it did in fact was to close down the cell-mediated arm of the immune system and they ended up with a recombinant virus which killed mice genetically resistant to mousepox even when they were immunized against the mousepox. This we think raises a big concern regarding the potential for similar modifications to smallpox. So that leads us then to our first question ...

Consistently offering such personalized renditions of the issues raised would have been likely to encourage a discussion much more centered on the moderators' thinking.

Rather than this, the exchanges were characterized by a complex dance between first- and third-person attributions. For instance, in an exchange that followed a slide reporting the oversight review mechanisms for experiments of concern recommended in the *Biotechnology Research in an Age of Terrorism* report, it was asked:

Excerpt 6.9

P20: Um, sorry, I was just wondering, kind of, how much of this is actually because people are genuinely concerned about the implications of the work that's going on, and how much of it is just so that (.4) the governments are just being seen to do something, in more of a sort of pandering to public opinion rather than genuinely because they think that there's (.8) an actual risk going on?

MD: So, so this is argument is (.) the politicians [think, think, yeah =

P20: [I'm just wondering whether that might be the case.]

MD: = yeah, but] but it's a good argument and it's one that we've heard in other places, that the politicians are looking at the possibility that =

P20: = They're just jumping on a bandwagon perhaps than genuinely.

(Seminar 9)

In this exchange, the evoking of third parties is part and parcel of a rather complicated discussion of motivations. P20 starts by proposing a binary question regarding the motivations of those involved with the oversight mechanism envisioned: is it motivated by genuine concern or a need to pander to public opinion? This question is not an overt attempt to get the moderators to offer their personal evaluation of the proposal outlined. Yet, it is certainly possible to envision how an answer might not only have stated the reasoning of those involved, but also evaluated the appropriateness of this reasoning. MD instead responded with a formulation of P20's question, one which selectively picked up on her reference to 'government' in the second half of the binary posed.

Before he could finish this formulation, P20 interjects with 'Well I'm just wondering whether that might be the case'. While a seemingly fairly inconsequential statement, it put the question back to MD regarding the motives for the activities. It would have sounded rather odd or at least conversationally stifling if MD at this stage had responded with yet a further formulation of P20's more qualified and speculative statement. In addition, P20's adding of the qualifying phrase that she was 'just wondering' brings further difficulties with asking a probing question about why she thought the binaries posed was appropriate.

MD's response did not adjudicate between the alterative accounts of motivations proposed. Rather it itself referred to statements made by third parties in other seminars. In doing so MD was able to acknowledge P20's contribution but also avoid ascribing any motives to those engaged in the oversight deliberation process. In the exchange no special emphasis was placed on 'good'. The follow on interjection by P20 gave a qualified support to one option proposed.

The evoking of the third parties was potentially open for many lines of questioning. The reference to other places in Excerpt 6.9 or 'people' in the IL-4 mousepox introduction noted above might have been queried by participants regarding why the moderators were bringing up such points at the times we did. These might have been interpreted as attempts to shore up a certain evaluation. As well, participants might have vocally judged the citing of third-party statements as insufficient. So, P20 in the Excerpt 6.9 might have followed up any of the third-person references by asking what the moderators thought about the issues. Yet, at no time were the motivations for our citing third parties queried in such ways and, as noted previously, our motivation and agendas were basically non-issues during the seminars. As Clayman and Heritage argued in relation to news interviews, the extent to which the evoking of third-party statements are orientated to as simply neutral acts should be understood as a 'collaborative construction' – this in the sense that both interviewer and interviewee must act so as to not draw attention to the contingencies of what was said and how. Much the same could be said of a notional sense of neutrality fostered in our discussions.

As mentioned in the second section of this chapter, we attempted to make our thinking and choices topics for discussion. That included the use of the aforementioned feedback slide. It also included finishing a subset of the seminars with self-initiated statements of our thinking about what 'the problem is' and what 'needs to be done'. Yet in the four seminars this was done, only one resulted in any discussion. In one there was no response at all and in another a minor point of disagreement was raised. The fourth consisted of what the author would regard as the only overtly hostile exchange between a moderator and a participant.

Discussion

This chapter has provided an analysis of the dual-use seminars in relation to themes associated with bias and neutrality. It noted something of the contrasting ways notions of intent and motivations figured in discussions about dual-use research and the partiality of actions: they were both pervasive in accounting for (certain) individual's actions and also judged effectively irrelevant. A variety of moves could have been undertaken in response to this situation. It would be possible to use the cumulative responses to document the views of scientists or to attend to the detailed discursive features of language. Instead, this chapter attended to the strategies employed to make the reasoning underlying, and the implications following from, accounts topics of discussion.

While the seminars sought to promote and did indeed promote animated questioning, the extent of this was bounded in significant respects. Although in the abstract university seminars might be regarded as places characterized by free-flowing and open critical dialogue, this did not take place in practice. Despite the considerable scope for questioning our motives or contributions, participants rarely did so. This may well have derived, at least in part, from certain common interactional practices associated with questioning. The very question-answer format of discussion as well as speaking on behalf of third parties arguably reduced the prevalence of certain types of challenges.

From a practical point of view such a lack of questioning had certain immediate interactional benefits. We as moderators were able to get on with a line of questioning without justifying the exact reasons for everything that was said. It was possible to plan the timing and content of questions to explore our emerging sense of prevalent responses. It was also possible to get to the end of such questioning without overt antagonism. Yet, in key respects, this situation had severe downsides. Despite the attention given to publicly testing reasoning, the overall exchanges fell far short of the Model II forms of learning detailed in Chapter 3.

It seems quite likely that the exchanges entailed the suppression of negative evaluations and the use of covert attributions on the part of participants regarding the moderators' action. It would be rather odd if the emphasis placed on the motives and intent of scientists and politicians did not extend to the moderators.

A different form of inquiry would have been to combine questioning with advocacy. Each topic could have been introduced by a thoroughly personal appraisal of its significance and our assessment of the question. This would have undoubtedly generated conversations with a different center of gravity for debate. However, as noted, attempts by us to explicitly introduce certain types of evaluations or elements of advocacy floundered. Why doing so in a more thoroughgoing manner would have resulted in anything else was not obvious. Making a change after a number of the standard seminars were completed would have meant departing from a seminar format that by and large fostered lively and, in significant respects, probing conversation. Such are the choices made in social intervention.

7
Expertise and Equality

Who decides and how are perennial concerns in public policy. In relation to many matters, the answers given will depend on how the issues at stake are defined. The greater the extent to which any definition portrays them as technical ones of calculating how best to achieve some agreed end, then the more likely the who and the how of decision-making will be restricted to the activities of specialized experts. The greater the extent to which any definition suggests the issues involve contentious ethical choices, then the more likely the who and the how will include some form of involvement of 'the public'. In democratic countries, most adults are treated as equally morally competent and they have expectations that decisions are taken in a manner that reflects the overall will of the populace. Therefore, just how concerns about expertise and equality play out for any given area is a matter of some significance.

In relation to the dual-use issues of concern in this book, previous chapters have contended that figuring out what should be done is a complex and vexing task, not least because what should count as the issues at stake is itself often in doubt. Against this background, though, it has been noted that the policy call for research community self-governance is a recurring one. The general risk/benefit assessment procedure central to many such calls exemplifies this preference (see Chapter 2). Therein, experts are meant to weigh up complicated scientific, security, and ethical considerations in order to determine the suitability of a research project or publication. The specific ways in which this general prescription has been formulated often suggest that a number of presumptions underlie these assessments: determinations of risks and benefits are scientific matters, a non-*ad hoc* calculus can be devised for assessing them and, through this, widely replicable assessments of the greater good can be made. Such a way of thinking contrasts with others, such as treating the weighing of risks and benefits as a largely political process where social groups are likely to have sharply opposing interpretations of what should be done.

While accepting that the need for community self-governance gives a center of gravity to decision-making, it does not resolve questions about exactly who should decide and how. If scientific experts are to be largely left alone to assess what needs doing, the question then becomes 'who are these experts?'. So do all those called 'scientists' or 'biologists' equally count as experts? Presumably not, given their varying training and competences. Who then is to decide who is expert among the experts?

Of course, there may be those who do not accept community self-governance as sufficient. If war in the twentieth century became too important to be left to the generals, research in the twenty-first century might be too important to be left to the scientists. Along these lines it might be argued that the uncertainties and unknowns entailed, and the stakes involved, mean that political or public involvement is required.[1] Scientists and other highly specialized experts have been criticized in the past for having a blinkered mind-set that means vital social concerns are rendered invisible to them.[2] If only to prevent critical voices from becoming too prominent, those advocating an expert-based system need to consider how it would relate to wider societal deliberations.

This chapter examines the themes of equality and expertise in the seminars. As with the previous chapters in Part II, this one recounts the appraisals given in relation to the changing strategies we employed as moderators to inquire about this matter. As with those previous chapters, this one also assesses the parallels between the seminars' substantive themes and their interactional dynamics. When one moves from national policy decision-making to group-based discussions, the dilemmas associated with how expertise and equality relate do not disappear. Certainly, that those seminars took place in a university setting where the intellectual worth of ideas is notionally crucial did not immunize them to awkward tensions about who should speak.

Dual-use research in the public domain

Concerns about equality and expertise were voiced in relation to a variety of topics. One area was the involvement by the public or political representatives in oversight systems. Although it was uncommon that statements were made in support of such involvement, six participants did express such sentiments. In relation to oversight this included suggestions that:

> I think my view would be that you certainly wouldn't want politicians doing it, but you might not want the scientists doing it either, although, because there are so many competing interests there about why scientists would want much more broad scope for their research than perhaps the public would. So I would say the scientific community initiate it, but not to leave it completely in the hands of the scientific community.
>
> (Seminar 3)

and that deciding about what research should go ahead is:

> not simply a scientific question, far from it it's an ethical and moral question and implicit within that ethical and moral question is the question of decision-making and in that context deliberately or participative decision-making in who gets the resources to do that and who has the ethical right to do that or not to do that?
>
> (Seminar 19)

Rather than surveying all of the topics where equality and expertise were pertinent though, this chapter examines one issue in detail: public communication. This has been a topic of discussion in recent years far beyond matters of dual-use research. To make certain educated lifestyle choices and take positions on many topics requires that members of the public have some grasp of science. That much of the research conducted in universities is paid for through public funding places a demand on scientists to make their work open and socially relevant. Yet, as indicated by disputes about the safety of genetically modified foods, vaccines, or tobacco products, just how and what should be told to the public can be a matter of some contention. The communication of science raises important questions about the place of experts in modern democracies.

Concern about what counts as appropriate communication is evident from reports such as the British Royal Society's *Science and the Public Interest: Communicating the Results of New Scientific Research to the Public*. It sought to provide practicing researchers with advice regarding whether and how to communicate their work. Central to the report's recommendations was the importance of serving the public interest, this being defined as communication that would:

- further the public's understanding of, and participation in, the debate of issues of the day;

- facilitate accountability and transparency of researchers, their funders, and their employers;
- allow individuals to understand how the results of research affect their lives and, in some cases, assist individuals in making informed decisions in light of the results; and
- bring to light information affecting public well-being and safety.[3]

It also recognized, however, that the public interest had to be balanced against other considerations, such as national security. The report included a communication checklist for researchers that flagged concerns about the public interest implications of the results, their possible application, and the integrity of their findings.

While *Science and the Public Interest* identified a number of concerns that researchers should keep in mind, it left open how they ought to be balanced in practice. No attempt was made, for instance, to specify how national security considerations ought to be weighed against public interest ones. The report did, however, identify who should take such decisions – this primarily being individual scientists, where necessary in consultation with their peers. In this spirit, the Royal Society rejected calls for an independent agency to give advice on communication in favor of those in the research community becoming more conscientious.

Difficulties and disagreements in determining what ought to be communicated and how were rife in the British university dual-use seminars. Many of these alternative evaluations were informed by different models about the place of scientific expertise in society.

Before detailing the range of arguments offered, it should first be recalled that only over the course of a number of seminars did the extent of disagreement about public communication become evident. As noted in Chapter 4, we initially asked whether experimental results such as those involving IL-4 mousepox should have been widely circulated. The background for this referred to both publications in the *Journal of Virology* and the *New Scientist*. This was then followed up with a slide that suggested unidentified British researchers in the 1990s had come across concerns about the lethality of IL-4 insertion for pox viruses, but avoided deliberately raising concerns in the scientific community. Instead, they alerted government officials to the possible malign implications of this work. The slide was meant to get participants thinking in more subtle terms than whether to publish or not.

We, as moderators, only gradually realized that asking whether dual-use results should have been widely circulated conflated types of publications that were subject to quite divergent appraisals. Starting with the tenth seminar, we broke this question into two parts. The first one asked about the merits of publishing in the *Journal of Virology* and consistently led to overwhelmingly affirmative responses. In light of such statements, the second part asked about popularly communicating results through the *New Scientist.* The second consistently led to sharp disagreement.

Past cited exchanges (such as Excerpts 4.2 and 6.2) suggest something of the worries cited with communicating to a wide audience. These put limitations on 'the need to know'. One concern was that magazines such as the *New Scientist* often sensationalized stories in order to sell copies and in doing so they massaged the facts. This sensationalizing, not only raised doubts about the accuracy of accounts, but concern was also expressed that it could cause public fear that would then result in restrictions on research (e.g., Excerpt 4.4).

Various contrasting points centering on 'the public' and popular publishing were made during the seminars. For instance, after the comments by P3 in Excerpt 6.2 regarding concerns about popular communications and how scientists are unreasonably pressurized to make the most of their work, it was said that:

Excerpt 7.1

P2: But you don't think that you've risk, you risk it being (.) um (.4) publicized in a worse way if you're not involved that you have no influence on how, <u>how</u> it's put over [or] explained or whatever.

P3: [Well]

P3: I don't think you do 'cos I think ultimately you have the fall off position of the (.6) the journal, of the work that you've actually published.

BR: But, but, but would you say that, I mean, is that to suggest that it's a good thing to (.) to (.) to try (1.5) to do this publicity. I mean does follow from what you are saying or no?

P2: ((Inaudible)) to try and do it. I think (3.5) if if it's kind of, if it's going to be controversial it's almost inevitable that something is going to come out of it, it's probably better to be involved (1.6) um than to leave it to (.5) you know *The Sun* or something to

((Inaudible multiple voices))

P8: You have not got any influence on what is written.

(Seminar 19)

Subsequently this line of argument was followed up with the exchange that:

Excerpt 7.2

P4: I was going to say if you hold an argument that it is sensible to make the general science (.) um like the scientific community aware of what you are doing, does it not follow that it is sensible to make the um not the (.6) to make the general public aware of what you are doing too? Not just the science community, don't you also have a duty to tell the general public [about] um?

P9: [You do]

P3: You do but do you think that headline is a (2.2) [for instance is, is, is,] =

P4: [well do you think the] =

P3: = [informing the general public?]

P4: = [general public.] Do you think the general public would read it =

P3: [Yeah]

P4: = if it did not have that headline?

P10: ((Laughs))

P3: Well but then again what are we in the business of? Are we in the business of (.3) you know glamorizing (.3) um sexing up um you know sensationalizing science? Is that what science is all about in terms of communicating to the public? I don't know.

(Seminar 19)

These exchanges echo themes similar to those in Excerpt 4.4 regarding whether scientists should take the lead in pre-empting any popular coverage, the possible effects of any coverage on public perceptions, the relative competences of individuals to receive and create news accounts, and whether it is unavoidable that results get out.

To publish or not to publish in the popular media?

In broad terms, it is possible to identify two options regarding the communication of the implications of dual-use research: one that the public should be told about such findings and, second, that communication should be restricted to those with some claims to scientific expertise. These evaluations were offered in comparably equal frequency over the course of the seminars. A key point, though, is that these two alternatives were justified by varied lines of thinking regarding expertise and democracy.

As in Excerpt 7.2, some participants suggested the public has the right or need to know. Such sentiments were expressed by one of the Australians taking part in the IL-4 mousepox experiment who was quoted in the *New Scientist* article 'Disaster in the Making' as stating, 'We wanted to warn the general population that this potentially dangerous technology is available.'[4] Seminar participants identified a range of reasons for fostering greater public awareness through popular communications. These could 'serve as a warning' and they were required because scientists 'have to be accountable to the people who pay them, which is generally the public through their taxes.'

Expressed support for public communication was tempered in two important respects: first, in all but one of the seminars any backing for it by some took place along-side critical comments. Second, supportive comments were overwhelmingly tentative or qualified, especially so when probed.

In elaborated discussions, assessments about the desirability of public communication could be seen to turn on basic issues about how the implications of science should be handled:

Excerpt 7.3

P9: I think that one of the things about having something that says (.) One step away from (.) biological weapons would mean that (.) a lot of people would pay attention when previously they hadn't, and I don't mean scientists but more (.5) policy-makers, ethicists, philosophers, you know it's going to, people might think (.) if we're doing this we need to start looking at it and discussing it, whereas if it's. This would go unnoticed maybe by an awful lot of people apart from the scientific community. So maybe that's a plus for having it published in such a splashy way, because it (.) it then enables people to discuss things that they didn't even know were going on. So.

BR: OK, so it's a way to get a wider debate than [this] more ah closed model.

P9: [Hum]

P10: Also if you start to be selective about domains that you publish in you (0.5) you start to tighten assumptions about specific people and specific uses in those domains. And unless you have a policy of open publishing, you, those assumptions are never going to be right. I mean, why just because it's published

in the scientific world that it's used for good rather than bad, who's to define what's good or bad? I mean.
(Seminar 6)

The comments made by P9 and P10 (both individuals with a professional interest in the policy aspects of science) suggest that popular communications can usefully encourage a broader range of attention to dual-use issues and that a wider discussion can counter the limitations of any assumptions that might otherwise be made.

The one seminar in which only (overall) supportive evaluations were expressed for popularly communicating, likewise, involved important issues about the nature of research and its place in society. A part of that seminar, for instance, went as follows:

Excerpt 7.4

P5: ... The way I read that is that it is saying we're one stage from being able to do something like this, shouldn't governments be (.5) looking at it (.4) and finding ways to defend ourselves against it if it actually does happen which you got to take account of ((inaudible)).

BR: So following on from the first bit, if we need to know then then people need to be as aware as possible so this is a (.7) this is a good way of doing it.

P5: I don't see a problem with it.

BR Yeah.

P6: It's also supposing journals like that ah want to sell lots of copies don't they. So it is more liberal with the language ((inaudible)) whether that is the true state of the art.

BR: But wh- what would that mean to this question about about whether they should go ahead and do that sort of (.4) publication? Would you have concerns then or?

P6: <u>Not</u> really because a lot of these (1.5) papers like *New Scientist* spice things up for the actual readership.

BR: OK.

P6: Because they want to sell copy, don't they?

BR: But that's not a <u>big</u> concern.

P6: I would not have thought so.

BR: I think there was (.) somebody, somebody at the back.

P7: The only concern I would have is if someone like the *Daily Mail* grabbed on this which is still a kind of semi-respectable scientific journal and then, y'know, dug in and came up with the sort of horror stories that people would kind of would react to. But in general I think forewarned is forearmed and back on the other points where you can't really keep information in a box. Once it has been done and you find something the best thing to do is to disseminate as widely and correctly and as accurately as possible because (1.0) if it's got negative applications some somebody somewhere will think up positive applications.

BR: OK yes was there?

P8: ((P3's)) point of earlier on about the duties or responsibilities of the editors and the peers (.5) is there, but sometimes the peers (.3) are looking at the science and not the wider aspects and by using general publications like the *New Scientist*, it makes editors and the peers aware of (.) potential (.5) uses that they never thought about before because they were thinking about the pure science in its own context.

BR: OK, it all seems rather positive then insofar that by saying in terms of, in terms of that sort of dissemination. Did you want to?

P9: Again I think that, um, y'know, that information is very important and you know and ((inaudible)) where do you draw the line at. At the end of the day if you are aware of what the risks are, or if the scientific community is aware of what the risks are, then obviously then you will be more alert to, if you see people doing things that might be slightly suspicious.

BR: OK.

P9: And that's probably, y'know, more protective than trying to hide the information and its consequences.

(Seminar 22)

Herein, several claims were made to support popularly publishing: P5 suggested it can act to spur government funding; P6 that the liberal language of magazines generally was not a problem and the knowledge would likely come out anyway; and P8 and P9 that scientists could better understand aspects of their own work through popular publications.

Yet, in those appraisals suggesting the need for expert-centered communication and decision-making, many of the arguments given for the benefits of public communication were reversed. So in a discussion about the slide detailing the low-key approach of British researchers working with IL-4, one participant said:

Excerpt 7.5

P8: I can see there's a need for public debate, but I also think that (2.5) well I am British of course, in, in, in a sense this is a better way to do it because the debate would be had by the people who need to have a debate, the people (1.0) un in authority who may design the antidote to biological warfare, and there is so much hysteria in the public about (.) weapons of mass destruction or not, um things that people think are going to happen in probably the United States or in Britain. I'm not sure that this adds to it or has a lot to do with. I think this is going to be taken by people who actually, probably know much more ...

(Seminar 3)

So while recognizing the need for public debate, P8 then goes on to suggest the importance of knowing about the issues at stake and limiting the discussion about what should be done to individuals with the proper competences. Echoing a similar assessment about the importance of expertise, in another seminar exchange where the merits of communicating through magazines such as the *New Scientist* was being discussed, it was said that:

Excerpt 7.6

P1: It's an interesting question. Of course, the snag is that there (3.3) the assumptions (.4) there are a lot of assumptions in this. Um I mean I'm sure you are probably aware that there's actually no, there's no real reason to suppose that if you did the same experiment in smallpox you would actually increase the virulence of the virus. Um (.5) it might do, it might not. It's (.3) there is a very open question. Um (3.5) I agree the issue of wider publicity is a tricky one. Um. And um having said that of course, the pub (.) the original publication probably would come to the (.) attention of the majority of people who had the (.) ability to actually repeat the experiment. This additional publicity was only coming to the attention of people who (.) actually have no facilities or technical know how.

BR: So, so is what you're saying (.) that people who (1.4) were in the know would have known about this experiment [((inaudible))]

P1: [Well, that's what I'm saying]

BR: So this was maybe =
P1: [Yeah]
BR: = a bit superfluous?
P1: Having said that I'm not very keen on (1.5) what I regard as (.5) well (1.3) yes
((Group laughter))
P1: sensationalization of what's really a more (.3) quite a fundamental piece of research that's being done. I don't like (.4) I don't like that kind of coverage.
MD: What do you see as the drawbacks to that, is it just that you've got a slight distaste for it or is it?
P1: No, I think it, I think it actually damages the perception of scientists in the public (.3) the public's perceptive view, I think it makes us sound like a bunch of cranks who don't know what we're doing.

(Seminar 14)

Herein, as in the previous excerpt, the issues brought to bear are limited to technical and scientific considerations. Given this framing, there is little need to question the place of technical experts or make dual-use discussions fairly open. Those experts that needed to know about experimental results would probably come to know about them through specialized publications. Because of the centrality attributed *expert* knowledge, the accuracy of any popular statements is a significant source of concern.[5]

Within arguments suggesting reasons for or against popular communication, there were certain premises repeatedly proposed to justify individuals' conclusions. As noted in previous chapters, the said inevitability of research frequently figured in responses. As noted in Chapter 6, though, inevitability steered assessments of popularly communicating in different ways. It could lead to treating this activity as a side non-issue or justifying communicating as widely as possible. Another key matter was the impact of popular communications on the likelihood of a bioattack. As in Excerpt 5.9, the headlines attached to dual-use research might or might not be seen to affect the prospect of would-be users becoming aware of it. Because of the manner in which such users were largely discussed as terrorists bent on manipulating fears, public communication might or might not also affect the desirability of employing bioweapons:

Excerpt 7.7

P7: I think it's a similar thing, if you talk about (.) how dangerous it is to put this information in the public domain then what (.)

kind of people of you talking about taking advantage of this being in the public domain? It's not (.3) going to be like the kids who will look up on the Internet and discover how to make a bomb and play in their bedrooms. You're essentially talking about labs that would be working (.) on making bioweapons anyway in which case as ((P2)) said they would be thinking about (.) doing this anyway themselves.

P8: The question is whether the wider publicity of this actually steers the direction of those people's thinking. If you are a terrorist =

P7: [Yeah]

P8: = wanting to do terrorist acts and you know that there is a hysteria ((inaudible))

P7: But you say you are a terrorist, you are like an independent terrorist?

((Group laughter))

P8: If you were a terrorist group, would you be influenced by what the world is saying [about that or not? I don't know.]

P7: [If I was terrorist group] that wanted to make a (.) modified smallpox I would be trawling PubMed journals ((inaudible))

P8: My question is would you choose to make a smallpox if there wasn't already a public hysteria about bioweapons?

P7: Yeah, because you would start to think what could I do with small pox to make it (.) more infectious. I'll add in this, I will add in that.

(Seminar 14)

In the first intervention by P2, additional dangers associated with popular communications are downplayed because of the demands associated with producing biological weapons. P10 draws attention to a prior question: where would that goal come from in the first place? As with contrasting overall appraisals of popular communications, much here depends on whether or not the issues at stake are portrayed as relatively narrow technical matters of achieving some end. Here again, as in many of the other excerpts, participants worked with various presumptions and models regarding the source of bioattacks, some (but no doubt only some) of which emerged through a process of dialogue.

The aforementioned exchanges illustrate the scope for alternative assessments and bases for assessments regarding the merits of bringing dual-use research to a public audience. So, divergent expressions were given for thinking about the appropriateness and outcomes of wide-ranging debate, the desirability of greater transparency about research, and what measures would ultimately serve public well-being and safety. In light of this diversity and disagreement about public communication, the insight provided by suggesting researchers should attend to 'the public interest' to reduce contention about such communication would seem rather limited. Just what this is and how it can be achieved in the area of dual-use research are matters of widespread contrasting thinking.

Expertise in interaction

The previous section illustrated the contrasting space given for expertise in assessments about what and how to communicate life science research. As illustrated, much scope exists for dispute about what is in the 'public interest'. Especially in light of the disagreement evident about the wisdom of popularly publishing, the question can be asked about how participants shored up the credibility of their own claims or disputed others' arguments.

Such a question is one of many that might be asked by way of attending to how expertise and equality entered into the interactional dynamics of the seminars. While faculty seminars in universities might be portrayed as places of open intellectual inquiry, not all voices in them necessarily speak with equal authority. Tracy identified a number of routine problems of intellectual talk in the communication department seminars she examined:

1a. Ideas are not challenged because the presenter possesses high institutional status.
1b. Ideas are criticized without recognition that the presenter is an inexperienced idea presenter/developer.
2a. Intellectual 'discussion' becomes long-winded monologues displaying various participants' knowledgeability.
2b. 'Intellectual' discussion becomes fast paced, lively expression of ignorance and ill-informed opinions.[6]

Intellectual talk in such settings – if it is to be judged as informed and appropriately questioning – must find some way of reconciling such concerns. The rest of this chapter considers how expertise and equality were consequential for interaction in the dual-use life science seminars.

To begin this, it can be noted that scientific expertise was a widely accepted basis for commentary. Against the wide-ranging political, security, and policy matters raised, no one said anything along the lines of 'I cannot answer that, I am only a scientist' or 'I am a scientist, I am not sure if I am qualified to speak on that topic'.[7] The questions addressed were never identified as matters that other professionals would have to settle (for instance, politicians, ethicists, or policy analysts). The only four times any participants qualified their responses by noting the limits of their general profession was when they identified themselves as non-scientists.

Yet despite the unquestioned place of scientific expertise in general, some individuals were singled out because of their specialized expertise. As mentioned in Chapter 5, for instance, in three of the four rare occasions in which one participant selected another to answer a question, this was done through explicitly noting the second's expertise in a particular area of science.

In general, between participants, expertise was drawn upon to *affirm* rather than *negate* the value of contributions. There was no instance of one participant critiquing the statements made by another through citing that person's lack of professional credentials. So, for example, no one said anything to the effect that 'You are an ecologist, not a virologist, so I think we should listen to Professor V'. Although it is not possible to know how many individuals from outside the life sciences attended the sessions or who in the audience knew of their involvement, no one ever commented that such individuals were non-scientists and therefore had a limited basis on which to speak.

No one overtly questioned the expertise of presenters either, even though we were regularly introduced as a sociologist and a former neurophysiologist. Instead, criticisms of our knowledge, if they were indeed criticisms, were done in a rather indirect fashion. So, as part of asking whether the IL-4 mousepox experiment should have been published in the scientific press, Malcolm Dando said that we were 'much nearer dangerous things here than' in the previously discussed case of the artificial synthesis of poliovirus. This point was picked up in the response that:

Excerpt 7.8

P3: When you say much nearer, is (.3) I don't think you necessarily are, perhaps you are. I could see perhaps a non-scientist would perhaps seek to argue that you are ((inaudible)), but mousepox as opposed to smallpox is different because smallpox is quite tightly regulated as to how you get a hold of it. I suspect that mousepox virus (.4) isn't and it might not necessarily, IL4 might not work the same way in smallpox as it does in mice. I don't think the problem is again (.) publishing the results of the experiment, but it is some way to control (.8) people getting hold of lethal strains of human pathogens such as smallpox.

MD: So you think (1.5) if I understand you correctly, that there's been too much of a pantomime made by non-scientists about this. It's difficult to get hold of smallpox, if it is at all possible, and anyway (.) putting human IL4 into smallpox might be a very different technical business than sticking IL4 in mousepox.

P3: Yeah.

(Seminar 3)

In this exchange, rather than confronting the statement by MD directly by questioning of whether he really understood the facts of the matter, P3 used the third-person footing (see Chapter 6) of 'non-scientists' as part of a rather hedged assertion that 'perhaps a non-scientist would perhaps seek to argue that you are' nearer. Given the previous discussion and introduction to the seminar, it would seem somewhat unlikely P3 is implying that MD is a non-scientist. Nothing in the tone of P3 suggests her comments were hostile. In any case, MD orientates to her response as directed at unidentified others through his own use of the third-person category of non-scientists.

In contrast, however, the appropriateness of the expertise of those outside of the seminar was often questioned. This applied to 'the public' at large as well as specific individuals in that public. A colorful example of this is given in responses to a slide that included a quote from a report of the UK House of Commons Science and Technology Committee that 'If the scientific community does not take stronger action to regulate itself then it risks having ill-judged restrictions placed on it by politicians.'[8] Among the responses included the contention that:

Excerpt 7.9

P2: I can't believe, I can't believe that statement has been made. I can't be (.4) Tony Blair, George Bush, George Bush is so afraid of science.

((Laughter and inaudible talk))

P1: Tony Blair isn't much better really on science is he?

P2: ((P1)), I mean Tony Blair is one, ah, ah, that has stated about science that he knows nothing about it, and now they're going to bring controls in, it's just (.) absolutely, I just cannot [understand it.]

P29: [Wa, wa] in reality this is the situation for the last 180 years. Society as a whole at large seizes on something when it when it feels under threat, then it turns against that following a period of years after that. Controls dropped to the point where it becomes reasonable and a bit technical, but this is the way our society works.

((Ten lines omitted))

P2: The quangos set up by all these worthy people, well tell me, you know, Mister Politician, what's your expertise in these,well, I went to a science lesson once in school so now I'm an expert. I, I want to know how legislation works on this. I mean the House of Commons Science and Technology committee would have credibility if they had scientists and technologists on it who were (.) qualified, skilled, and experienced. I'd listened to them of them and they're not. They're lawyers, they're art graduates and stuff like this. Nothing (.4) no harm. It is like me trying to regulate <u>art</u>, if you like, and publications and, and this sort of thing.

(Seminar 4)

While participants brought a variety of expert knowledge to bear in their responses, what relatively few did was to bring in what might be referred to as 'dual-use specific information' as part of the discussion. Only 19 individuals made reference to information beyond that provided by the presenters about dual-use policy developments, the past history of biological weapons, biosecurity events (e.g., efforts to reconstruct 1918 Spanish Flu), or the provisions of arms control. Three of these individuals' claims were inaccurate. Almost all of these were passing in character; such as indicating basic knowledge about previous state bioweapons programs or noting some awareness with policy initiatives. No participant response ever attempted to correct inaccuracies in the basic slide information we provided.

Thus, in terms of the interactions in the seminars, while participants were experts in their particular fields of the life science, few displayed knowledge about the policy matters under question. Our role as *initiators* of a conversation that few individuals displayed knowledge about meant we could not take refuge in some of the prescriptions for facilitating focus groups and other forms of social inquiry. So as presenters we could hardly pretend not to have expertise in the topics under discussion, as is often advised.[9]

The points in the last two paragraphs underline the dynamic negotiation of expertise and equality in the seminar interactions. If the participants had repeatedly displayed knowledge about the dual-use policy issues at stake, the exchanges could have entailed quite detailed and thorough examinations of particular issues. Instead, overall they were characterized by a division in expertise. Participants indicated expertise in their field of study, while the moderators were the ones that marshaled information about the policy issues under examination. As such, the quotes, figures, and information we provided became highly important in establishing the terms of the discussion. Our by and large overall control of the questioning process (see Chapter 5), combined with their comparative unfamiliarity with dual-use policy matters, meant our interventions as moderators could be highly consequential for the types of exchanges generated. This was the case for both the pre-scripted monologues given for each slide as well as the unscripted dialogue that followed. The tension faced throughout the seminars was how to introduce a topic to initiate discussion, but not do so in a manner that then closed it down.

Comparable tensions are arguably pervasive in attempts to generate thinking about matters where the audience has restricted knowledge of the issues at hand. For instance, this book has sought to promote reflection regarding a range of issues associated with dual-use research and social inquiry. In doing so I, as the author, have sought to refrain from simply offering an account of what 'really happened' in the seminars in order to instead be attentive to what Dewey referred to as 'the condition of the uncertain' in the process of inquiry. This has been done in part to promote further questioning about research practices and methods. Harking back to the points made in earlier chapters about the topic/resources distinction, this questioning has taken place only in certain ways. Some issues have been taken up whilst others were left to the side. As the reader, you have brought in your own considerations to bear which have suggested many additional lines of questioning, even if you have (at best) a highly limited first-hand experience with the seminars. A substantial difficulty is writing a book meant to aid learning and promoting experimentation is gauging what sort of analysis to provide.

A value and challenge of discussion-based seminars (as opposed to reading a book) is that the private reflections of individuals can become matters for group interaction. As such, key issues are how and when those initiating discussions seek to marshal claims in the pursuit of encouraging dialogue.

Strategies for interaction

This section recounts the strategies we employed as moderators in negotiating expertise and equality. In initiating this work we were uncertain regarding how claims to expertise might factor into the interactions. As will be recalled from Chapter 3, based on prior one-to-one interviews undertaken with neuroscientists, a danger perceived at the time of commencing the seminars was that a division in expertise might result in the discussion descending down two unfortunate paths: a one way 'grilling' of them on our part or a narrow technical lecture from them at us. Yet, given the number of scientists assembled in one room and the attention given to dual-use issues in the popular and scientific communities since the 2003 interviews, we wondered whether this past interview experience would prove a reliable guide. That the seminars largely consisting of varied contentious cases in which scientists or scientific organizations were taking the issues seriously was meant to minimize the prospect for participants simply lecturing the moderators.

In relation to matters of strategy, to begin with it can be noted that just what was said in scripted moderator monologues developed over time. The unscripted moderator interventions were highly dependent on the specifics of what was said. It is possible, however, to make some general comments which indicate the broad orientation we eventually adopted in thinking about what information to cite and how. After presenting the information related to each slide we sought evaluations about the questions asked and, in line with Argyris's suggestion (see Chapter 3), then further questioned in an effort to make explicit the assumptions and inferences informing those evaluations. If participants asked for further details related to the policy issues as part of the discussion then we provided them, but otherwise we tried to refrain from bringing in additional information or our own evaluations after posing the question on each slide. The aim was to cut off the provision of ever-more considerations in order to generate discussion about something. The details included within each initial slide could then be revised as we became aware of recurring queries.

As few participants showed any familiarity with the dual-use issues at hand, in general we sought to question the basis of their statements rather than forward further statements of our own. Consider, for instance, the following excerpt:

Excerpt 7.10

P9: I mean the term biodefense is actually pretty <u>off</u>ensive if you think about it and no pun intended. I mean to do <u>bio</u>defense research presumably means that one has to conduct bio <u>off</u>ence research in order to actually (2.0) combat the agents that you are actually concerned about. Is that not the case? I mean it strikes me that.

MD: It is a difficult distinction, but I think you can (0.6) you can make a reasonable distinction. You can say (2.0) if, if there (.) is a clearly sort of understood threat, a threat agent out there which is clearly known to be a threat agent. Then doing work =

P9: [Um hum]

MD: = on it might be justified in terms on biodefense.

P9: Um hum

MD: But if you actually imagine a threat and then work on that imagined threat and then you are into sort of a different kind of sort of mirror imaging of threats =

P9: [um]

MD: = and it may be possible to keep those two things apart.

P9: OK.

MD: Maybe.

BR: But your, your concern here is that there's a (.) implicit, almost within this there is this element of offence which does raise a concern?

P9: Well I think if the idea is to, is to keep ahead of your potential, perceived or or unseen enemy (1.0) um then by in so doing you are upping the ante, are you not? I mean it's like if you, you looked at the global defense industry. Um if that were to cease to be, um, you know um ultimately I think you would find the net result would be that there would be less conflict. So I mean just turning it round (.5) yeah I mean seeking to run faster (.) well the danger is that sometimes in trying to solve (.) perceived threats you can actually create much worse problems in the first place I think.

(Seminar 20)

Here P9 starts by offering an evaluation and a question about the interrelation between biodefense and bio-offence. MD followed this up with a statement regarding how such a distinction could be made, to which P9 eventually said 'OK'. While at one level a useful exchange, the statement by MD was rather limited in its potential to generate further consideration about the reasoning for P9's statement. Perhaps if P9 or others in the seminar indicated a significant familiarity with biodefense or dual-use issues generally, MD's statement could have been followed by a counter-argument or agreement which then could have led to lively further discussions. However, as a matter of routine this did not happen. For us as moderators to offer evaluations regarding matters of dual use, even when asked to speak to such topics, was generally to stop a line of conversation. As such, the type of intervention made by MD in Excerpt 7.10 is something we tried to avoid if possible. When directly asked to comment on a question, we tried to offer a statement but combine that with a probing question back to the respondent. The subsequent question forwarded by BR in Excerpt 7.10 was motivated by an attempt to allow a space for P9 to raise his underlying concerns with the audience. Realizing this overall strategy took some effort to do though.

The plan to question the basis for their evaluations rather than forward our own had its problems. Especially given the time constraints of the sessions, whether or not and just how we should correct inaccuracies we perceived in their assumptions was a continuing source of concern (e.g., regarding the competences of British parliamentary committees as in Excerpt 7.9 or the terms of the BTWC). A danger was that frequently interjecting about the facts of this or that would move the interactions closer to a one-way provision of information.

The issues in the last few paragraphs point to something of the trade-offs and tensions in determining how claims to expertise should enter into discussions. These comments, though, were made in relation to the limited display of knowledge by participants regarding the substantive matters raised. However, at the other extreme, many participants could claim expertise in the scientific areas under question. While participation was open to everyone equally in so much as almost no overt attempts were made to dispute anyone's basis for speaking, some individuals spoke more authoritatively than others. In perhaps the most extreme instance of this, in response to the first question of the seminar regarding whether artificial synthesis of poliovirus should have been done, it was said that:

Excerpt 7.11

P1: Eckard Wimmer is a bit of an attention grabber, he's a damn good virologist, but an attention grabber. But what he did was brought to the fore, um (1.0) back in the sort of early 1990s there was an International Congress of Virology when (.5) the first discussion started taking place about destroying the last stocks of polio or the last stocks of smallpox. And the people were saying there's no point destroying it because um we may need to come back and study it at some stage. And back in the early 1990s it was agreed back then that doing this sort of experiment was absolutely child's play. Based on the known sequences you could resynthesize the virus as and when we needed to. So I'm surprised that it took this long, I mean, I saw an interview with Wimmer and he basically said that the reason why it took them so long was Friday afternoon experiments, they didn't want to rush it. Um (.5) so (1.0) should it have been done? Yes, it brought at long last the attention to people in government that these sorts of experiments are pretty easy, if you've got the time, the effort, and the money, there's nothing to stop you.

MD: So if I understand you correctly, this should have been done because it was necessary to bring this to the attention of government?

P1: Yeah, people people in power just didn't realize how easy it is to do some of these experiments in science these days.

MD: And (.) the reason they needed to be brought to their attention was because of the dangers involved? Or =

P1: = Well basically really ((inaudible)), just because we've destroyed all stocks of any particular virus, be it smallpox or polio, doesn't mean it's gone forever. Based on the fact that we have sequences you can now, it's a bit like saying we're going to destroy all nuclear weapons (.) off, off the face of the earth. As long as you still have the knowledge there you can still re-create it sometime in the future. So it's exactly the same for a virus. Because of the sequence is out there, the ability to resynthesize these viruses is very, very easy, but it hadn't got through to these people. They still thought that because smallpox ah stocks had been destroyed, ah, um and polio there was only a few licensed holders around the world, well then we were all safe, and huge amounts of protection and double doors and everything else. Whereas these such experiments illustrate that if you are a reasonably well-equipped lab you can synthesize polio or smallpox very, very easily.

(Seminar 5)

Here a senior faculty member combined claims to scientific proficiency with a smooth display of substantive knowledge about dual-use policy issues. In this and a few other cases where senior faulty members gave lengthy and considered responses to the first question, either no other participants offered responses or those that did only issued short echoes of support. All of these types of statements gave strong support to publishing or conducting the experiment in question. While it was possible to ask for further elaboration of such statements, a danger of such authoritative accounts was that the subsequent seminars interactions would squarely center on the contribution from a certain individual. We made the decision that this was not an outcome we as moderators would welcome, even if it accurately reflected the distribution of familiarity in the audience with the issues at hand.

As moderators we dealt with this perceived danger of dominance through revising the content and sequences of questions posed as outlined in the second section of this chapter and in Chapter 4. The aim was to find ways of probing our emerging sense of likely sources of contrasting responses by participants, particularly the categorical support for publishing or conducting certain experiments. So, subsequent to the exchange in Excerpt 7.11, we are able to ask questions about the public's need to know and the advisability of the significant allocation of public funds in biodefense programs. Since such topics were likely to spark lively discussion, this would test the logic of just what 'needing to know' meant. That test, though, was built into the design of the seminar, rather than being one that would depend on a highly situated challenging of certain statements. Through this, any initial categorical and definitive statements could then move from being a threat to discussion, to part of the material for generating discussion. The manner in which the subsequent disagreement contrasted with previous unanimity would underline that the matter of dual-use research required considered thought.

Final remarks

This chapter began by highlighting the importance of who decides and how in matters of policy. In modern democracies, such questions raise crucial issues about how claims to expertise and equality mix in deciding what should be done. The seminars consisted of contrasting substantive talk about the need for those beyond scientific experts to be part of debates about dual-use issues and entailed a negotiation of expertise and equality in their interactions.

In relation to the substantive discussion in the seminars, this chapter focused on debates about the merits of public communications. The communication of the dual-use implications of research to the public raises basic issues regarding who needs to know about possible concerns and how they should be informed. As argued, the merit of public communications was a topic that repeatedly generated disagreement between participants. Alternative appraisals of whether dual-use concerns should be raised beyond the scientific community were justified by employing different ways of thinking about the place of expertise and equality in society. Some of the difficult issues on which the evaluations turned in these seminars included: Should scientists partake in communications activities whose ultimate message is outside of their control? Are technical experts best placed to judge future possibilities or does the very specialization that underlies their expertise limit their understanding? Is it necessary to 'hype up' issues to generate public concern and, if so, is this a problem? Who should determine what hyping up means? Does drawing widespread attention to possible threats make them more likely or would this merely alert those who were not able to carry out threats? Such tension-ridden issues were raised alongside the types of concerns noted in Chapter 6 regarding whether those in universities should seek to promote a wider 'appreciation' of their work or whether this was antithetical to their role as knowledge generators. Given the ways such thorny questions were brought to bear, there is no easy recourse to notions of what is in the public interest to resolve disputes.

In examining the tensions and dilemmas associated with expertise and equality in the seminar discussions, this chapter has also conveyed a sense of the choices made in deciding how to initiate a conversation about these issues. Some matters were raised as a topic for discussion in certain ways and at particular points in time. Some types of participant and moderator interventions had to be cut off to enable others. The appropriateness of the choices made – and particularly whether they promoted the learning objective sought – are matters open for debate if not always matters openly debated.

Part III

The Limits of Method

8
Closing Remarks

Biotechnology, Security and the Search for Limits has sought to promote questioning about the place of research in contemporary public issues. Against the somewhat unprecedented and somewhat re-emerging concerns about the 'dual-use' potential of life science research, it has considered what might be done in response. As has been contended from the start, though, assessments of what needs doing are inescapably bound up with how an understanding is developed of the issues at stake. As such, any analysis of the limits, utility, and like matters of the life sciences in relation to dual-use concerns should be undertaken alongside an analysis of the limits, utility, and the like of the inquiry carried out into this topic. As part of such a process, throughout this book attention has been given to the selective choices and challenges associated with what becomes taken as topic and resource in analysis (as in Chapters 1 and 2). Depending on the goal of analysis, accounts by scientists and others about the dual-use potential of research might be taken as evidence for making claims about the world, items for study in their own right, or some mix of both.

Part II of *Biotechnology, Security and the Search for Limits* considered an experimental line of empirical inquiry in relation to these substantive and interactional issues. In this, Malcolm Dando and the author turned existing departmental research seminar series in British universities into forums for generating dialogue about dual-use research. As argued, at one level the seminars entailed both discussions about openness and closure, neutrality and bias, and expertise and equality; and at another level the dynamics of the conversations themselves could valuably be understood in relation to these themes. Indeed, the further point can be made that the account given in this book of the substantive and interactional themes of the seminars itself entails a struggle for and against openness, neutrality, and expertise.

Both in the doing and in the telling of these outreach seminars, the attempt was made to find ways of acting that promoted mutual learning. So, in deciding what got brought to the foreground to be questioned and what remained in the background, the aim was to find ways of probing participants' thinking to foster further deliberation. Yet, such a broad statement of an aim only provides the thinnest of guides to the interactional complexities of how the conversations were planned and evolved.

Chapter 3 provided a description of the underlying pragmatic orientation that informed the preparation for seminars. Central to that orientation was the desire to maintain a sense of the conditions of uncertainty of inquiry in the quest for greater certainty through inquiry. As such, the certainty sought was not some solid ground of truth for all time for any purposes. Rather, it was a certainty of understanding – a wisdom it might be said – of how to promote the practical goal of having conversations about dual-use issues.

As a result, the seminars undertaken were an attempt to couple an empirical research agenda with an emerging sense of the substantive issues associated with dual-use research. This was elaborated in Chapter 4, which recounted how the focus group type sessions held with faculty members and students transformed as our sense of likely responses emerged. Through gaining a sense of responses and the dynamics of the conversations, we sought to subject the assumptions and inferences informing participants' evaluations to scrutiny. Topics, information, and questions were marshaled by us as moderators so as to probe dominant responses and enable these to become matters for deliberation. While this overall strategy facilitated discussion, it also limited it as well.

So, Chapter 5 examined the substantive and interactional themes of openness and closure in the seminars. As argued, when asked whether there were experiments that should not be done or some research better left unpublished or otherwise restricted in dissemination, participants overwhelmingly responded in ways which could be characterized as stressing that 'we need to know' and 'we need to do the research'. With a sense of expectations for responses, we as moderators were then able to ask follow-on questions that posed issues such as: 'So if "we need to know" then who is the "we" that should know, just how do we get people to know, and just how much do we need to know?' Through the contrasting responses generated through such follow-on questioning, differences in the assumptions and inferences informing the notion of 'needing to know' could become matters for joint examination.

Such a plan to generate substantive discussion could not have succeeded though, were it not for certain patterned interactional dynamics. For instance, if participants repeatedly put questions back to us to gauge our thinking or questioned the framing we gave to topics, these would have disrupted the structured planned question and answer format adopted. If the pattern of interactions between sessions turned out to be highly variable, then it would have been quite difficult to prepare. As contended in Chapter 5, at an interactional level, the seminars were both remarkably consistent and civil. The conversations generated were likened in terms of their dynamics to something of a cross between school teaching and news interviews. As with both, by and large we as moderators were the ones asking the questions and the participants responded to what was posed. The particular attempts made to promote further questioning by them of us foundered. While the interactions were more akin to news interviews than classroom teaching in the way we were *probing* reasoning rather than *assessing* individuals' knowledge; as with both there was an asymmetrical relation in what kinds of conversational turns were taken. So, we the moderators were almost never pressed to respond in some particular fashion. Through formulations and other techniques we regularly turned back the onus to participants to respond in ways that participants did not do to us.

In certain respects, the consistency across seminars was remarkable. The novelty of turning the typical departmental seminars into focus group type sessions and the lack of instructions to participants regarding the rules for proper conduct (as done in many focus groups) might lead some to think the dynamics of the interactional would have been unpredictable. Yet, we could count on a certain structure to the conversations. Participants, it might be said, 'played the game'. In other respects, that they played the game was unremarkable because many had been playing such a game of educational talk most of their lives. This is meant in the sense that the dynamics of the discussions shared much with the dynamics of educational instruction more generally.

One important way in which the seminars differed from typical focus group method prescriptions and news interviews standards was that as *initiators* of questioning about an out of the ordinary topic, we could not be said to be simply uninterested in or unknowledgeable about the topics posed. They were not matters we were researching about for a company or ones that we were addressing as part of our professional role of examining public issues of the day. As those actively bringing up a particular set of issues associated with dual-use research in a particular manner, there were many possible questions about our partiality.[1]

Yet, despite the scope for questioning in relation to issues of bias or neutrality, rarely did participants make our overall motives or the partiality of the information and questions posed topics for discussion. If they had, the conversations generated would have taken on a different substantive flavor and no doubt led to a different pattern of interaction. Moreover, they did not make each others' motives or partiality topics for conversation. What was much more frequent, though, was focus on the motives of scientists, politicians, and members of the public who were outside the seminar rooms as a basis for explaining what had been done.

A different pattern of interaction might well have been fostered if many of the participants displayed significant prior familiarity with the issues and initiatives under consideration. That they largely did not ,but that they did have specialized scientific expertise, meant that the conversations were characterized by a division of who said what between the moderators and the participants. This division contributed to an interactional dynamic that generated its own concerns. While we as moderators were able to present information and questions without these being constantly queried, how we initially framed the issues at stake became much more consequential than if participants were constantly bringing their own points to bear. The dilemmas and difficulties for us as moderators in knowing what to say paralleled those raised in the substantive discussion regarding the place of technical experts in democratic societies. Here, as with issues of openness and closure as well as neutrality and bias in science, we sought to design the seminars so as to bring underlying disagreements about expertise and equality to the fore to make them matters for further deliberation.

Learning about learning

By progressively and publicly questioning the predominant evaluations voiced through a self-critical experimental approach, the seminars discussed in this book were not only intended as a way of learning about participants' reasoning, but also as a way of learning about how to learn about that reasoning. As detailed, between different sessions alterations were made in what was said as well as how it got said. We as moderators sought to question how we went about questioning through this process of change. In the end, the seminars were brought into a form such that it was possible to generate fairly engaging conversations about dual-use issues.

And yet, the learning realized had its limits. Examining those restrictions is essential in considering what happened and what else could have happened. Chapter 3 considered the work of Chris Argyris to suggest the many ways in which learning was hampered in many interactions

and types of research. So, he has offered a contrast between prevalent Model I forms of learning indebted to defensive reasoning and the more desirable Model II forms. As developed in Chapter 3, these two models can be contrasted in the following abbreviated fashion:
This contrast begins to point to the many aspects of interaction and research identified in Chapter 3 that constrain learning.

Model I	**Model II**
• Offer un-substantiated attributions and evaluations	• Ensure reasoning is explicit and publicity test or agreement at each inferential stage
• Unilaterally assert evaluations	• Advocate position in combination with inquiry and public reflection
• Make covert attributions	• Publicly reflect on reactions and errors
• Protect inquiry from critical examination	• Inquire into your impact in learning

The interactions fostered through the seminars would receive a mixed assessment when compared to the types of highly demanding standards proposed by Argyris. While seeking ways to make reasoning explicit so as to put it up for deliberation was central to the seminars, on other counts the inquiry undertaken would not fair that well. To begin with, it can be noted that as the ones initiating the conversation and devising the seminars, we as moderators exercised considerable control over what happened. While we consulted with various members of the life science, policy, and other communities in undertaking the work, it could hardly have been said to be a form of equally mutual inquiry between us and the participants. The types of established long-term trust-rich commitments by all involved parties to inquiry required for Model II learning were not in place when each session commenced. The very manner in which the analysis of the seminars given in this chapter has leant itself to a language of 'us' and 'them' speaks to the limits of joint inquiry achieved.

In addition, while transforming the seminars over time on the basis of experience had the advantage of being able to more effectively probe predominate reasoning, it was the moderators that defined what 'effectively' meant. The revisions or options informing what was done in a given seminar were not available to participants, except when we as moderators chose to speak about them. Participants did get a sense of this in the feedback analysis sent to them afterwards, but again this was done by the moderators. There were other areas of concern regarding unilateral control. For instance, our ability to manage the turn-taking process meant we could and did choose to limit how much certain highly knowledgeable individuals contributed to the conversations. This could take place even if the extent of their participation was seen to accord with the distribution of familiarity in the audience with the issues at hand.

As suggested a number of times in this book, what was not said can be just as noteworthy as what was said. This applies to matters of learning as well. So Argyris suggested advocating positions in combination with public reflection so as to encourage feelings of vulnerability that could provide the basis for challenging personal thinking. Yet, within the seminars we only undertook modest efforts to advocate positions. What explicit attempts were made generally floundered in the sense that they brought a closing down of discussion. Also, being strategic about how we arranged the planned information and questions enabled us to confront expected responses without being directly confrontational (which might well have ended an exchange). Yet what this amounted to was an effort to minimize expressions of emotions (for instance, anger, shock) that might come from being more explicitly challenging; expressions that, if handled properly, could provide the basis for learning.

As a further limitation to be highlighted, Argyris has not only been concerned about interacting in a different way, but through this with changing what he called the taken-for-granted norms of inquiry. Such a radical move is said to be necessitated because of the deep-seated ingrained prominence of defensive thinking. This call applies as much to inquiry undertaken by those in universities as well as elsewhere. While these seminars challenged many of the typical prescriptions for departmental seminars and focus groups, in other respects they were quite conventional. So, in making little space for positive goals, in not seeking to develop practical skills for future learning in any substantial way, and in using conversational techniques such as third-person references (see Chapter 6), boundaries were placed on discussion. A different type of research design, say repeatedly working with a limited number of groups rather than 'one-off' conversations with a large number of groups, might well have enabled different types of learning by the participants and the moderators. However, this would have been done at the expense of the numbers of individuals that participated.

From the points above and those raised throughout Part II, it is apparent that equating academic discussion or 'intellectual talk' with open questioning and inquiry is mistaken, even when that discussion is notionally dedicated to such pursuits. Instead, for better and for worse, the limits of learning were also part and parcel of the basis for learning. The seminar design set out in the book 'worked' because of the conventions in the organization of interaction and it did not 'work' because of those conventions. In certain respects and in general terms, one or two people standing up in front of 30 brings with it a set of possibilities for progressing discussion (in terms of what is made a topic for deliberation versus a resource for argument) and managing interaction (as in turn-taking) that are not readily available when one individual interviews another. (Let alone when one reads the book of another.) The interaction between participants was essential in shifting the conversation away from a back and forth exchange between 'us' and 'them'. In the particular substantive case of dual-use issues, we were able to use the interactional dynamics and a sense of expected responses to repeatedly generate contrasting statements which could become topics for discussion.

A key question asked in the examination of dual-use life science research and of the interactions in seminars about it in this book has been 'what should be done?'. In considering responses, I have also wanted to examine what is entailed in asking this question in the first place. But as it is clear, all lines of questioning come to a close, if not a final end. A closing remark, then, I wish to say thanks to all those who contributed to the seminars. If wisdom consists, in part, of the ability to sustain a conversation, then I owe them much for what they have helped me to learn.

Notes

Introduction: What Should be Done?

1. J. K. Taubenberger, A. H. Reid, R. M. Lourens, R. Wang, G. Jin and T. G. Fanning. 2005. 'Characterization of the 1918 Influenza Virus Polymerase Genes', *Nature* 437 (6 October): 889–93.
2. T. M. Tumpey, C. F. Basler, P. V. Aguilar, H. Zeng, A. Solórzano, D. E. Swayne, N. J. Cox, Ja.M. Katz, J. K. Taubenberger, P. Palese, and A. García-Sastre. 2005. 'Characterization of the Reconstructed 1918 Spanish Influenza Pandemic Virus', *Science* 310, 5745 (7 October): 77–80.
3. G. Kolata. 2005. 'Deadly 1918 Epidemic Linked to Bird Flu, Scientists Say', *New York Times*, 5 October.
4. A. von Bubnoff. 2005. 'Special Report: The 1918 Flu Virus is Resurrected', *Nature* 437 (6 October): 794–5.
5. Ibid.
6. Ibid.
7. G. Kolata. 2005. 'Deadly 1918 Epidemic Linked to Bird Flu, Scientists Say', *New York Times*, 5 October.
8. R. Kurzweil and B. Joy. 2001. 'More Spanish Flu', *New York Times*, 17 October.
9. G. Kolata. 2005. 'Deadly 1918 Epidemic Linked to Bird Flu, Scientists Say', *New York Times*, 5 October.
10. For instance, see CNN. 2005. 'Researchers Reconstruct 1918 Virus', *CNN.News*, 5 October.
11. P. A. Sharp. 2005. '1918 Flu and Responsible Science' *Science* 310, 5745 (7 October).
12. D. Kennedy. 2005. 'Better Never Than Late', *Science* 310 (14 October): 195.
13. Ibid.
14. Ibid.
15. See as well, J. Kaiser. 2005. 'Resurrected Influenza Virus Yields Secrets of Deadly 1918 Pandemic', *Science* 310, 5745 (7 October).
16. BBC News. 2005. '1918 Killer Flu "Came from Birds" ', *BBC News*, 5 October; Reuters. 2005. 'US Scientists Resurrect 1918 Flu, Study Deadliness', *Reuters*, 5 October; and Associated Press. 2005. 'Researchers Reconstruct 1918 Virus', *Associated Press Statement*, 5 October.
17. I. Sample. 2005. 'Security Fears as Flu Virus that Killed 50 Million is Recreated', *The Guardian*, 6 October.
18. Associated Press. 2005. 'Researchers Reconstruct 1918 Virus', *Associated Press Statement*, 5 October.
19. E. Ghedin, N. A. Sengamalay, M. Shumway, J. Zaborsky, T. Feldblyum, V. Subbu, D. J. Spiro, J. Sitz, H. Koo, P. Bolotov, D. Dernovoy, T. Tatusova, Y. Bao, K. St. George, J. Taylor, D. J. Lipman, C. M. Fraser, J. K. Taubenberger and S. L. Salzberg. 2005. 'Large-Scale Sequencing of Human Influenza Reveals the Dynamic Nature of Viral Genome Evolution', *Nature* 437 (6 October): 1.
20. I. Sample. 2005. 'From frozen Alaska to the Lab', *The Guardian*, 6 October.

21. I. Sample. 2005. 'Security Fears as Flu Virus that Killed 50 Million is Recreated', *The Guardian*, 6 October.
22. The Sunshine Project. 2005. *Disease by Design: 1918 'Spanish' Flu Resurrection Creates Major Safety and Security Risks*, News Release, 5 October: 1.
23. The Sunshine Project. 2003. *Emerging Technologies: Genetic Engineering and Biological Weapons*, Sunshine Project Backgrounder #12 October.
24. The Sunshine Project. 2005. *Disease by Design*, News Release, 5 October.
25. Ibid.
26. J. Taubenberger, A. Reid, and T. Fanning. 2004. 'Capturing a Killer Flu Virus', *Scientific American* (January): 62–71.
27. M. Hopkin. 2004. 'Mice Unlock Mystery of Spanish Flu', *New Scientist*, 6 October.
28. J. Kaiser. 2005. 'Resurrected Influenza Virus Yields Secrets of Deadly 1918 Pandemic', *Science* 310, 5745 (7 October): 28.
29. Ibid.
30. C. Brahic. 2005. ' "Resurrected" 1918 Flu Virus Gives Insight into Bird Flu', *SciDev.Net* 5 October.
31. For other statements of 'hopes' and 'helps' rather than certainties and accomplishments, see BBC News. 2005. '1918 Killer Flu "Came from Birds" ', *BBC News*, 5 October; News-Medical.Net. 2005. 'Genome Researchers to Look at 1918 Spanish Flu', *News-Medical.Net*, 20 September, see: http://www.news-medical.net/?id=4905; and Reuters. 2005. 'US Scientists Resurrect 1918 Flu, Study Deadliness', *Reuters*, 5 October.
32. A. von Bubnoff. 2005. 'Special Report The 1918 Flu Virus is Resurrected', *Nature* 437: 794.
33. G. Kolata. 2005. 'Deadly 1918 Epidemic Linked to Bird Flu, Scientists Say', *New York Times*, 5 October.
34. A. Fauci and J. Gerberding. 2005. *Unmasking the 1918 Influenza Virus*, Press Statement, 5 October (Bethesda, MD: National Institute of Allergy and Infectious Diseases): 1.
35. Ibid. For another characterization of the risk as theoretical, see I. Sample. 2005. 'Security Fears as Flu Virus that Killed 50 Million is Recreated', *The Guardian*, 6 October.
36. D. MacKenzie. 2005. 'US Scientists Resurrect Deadly 1918 Flu', *New Scientist*, 5 October.
37. Ibid.; and Reuters. 2005. 'US Scientists Resurrect 1918 Flu, Study Deadliness', *Reuters*, 5 October. On the downplaying of concerns, also see A. von Bubnoff. 2005. 'Special Report: The 1918 Flu Virus is Resurrected', *Nature*, 437: 794–5.

Chapter 1 The Dilemmas of Dual-use Research

1. A. Jamrozik and L. Nocella. 1998. *The Sociology of Social Problems* (Cambridge: Cambridge University Press); J. Haydu. 1999. 'Counter Action Frames', *Social Problems* 46, 3: 313–31; and M. May, R. Page, and E. Brunsdon. 2001. *Understanding Social Problems* (London: Blackwell).
2. P. Conrad. 1997. 'Public Eyes and Private Genes' *Social Problems* 44, 2: 139–54, 159.
3. Including settings such as the National Academies and the Center for Strategic and International Studies. 2003. Meeting on National Security and

Research in the Life Sciences, 9 January (Washington, DC); and the National Academies. 2001. *Balancing National Security and Open Scientific Communication: Implications of September 11th for the Research University*, 13–14 December (Washington, DC); and the National Academies. 2002. *Report on Post-September 11 Scientific Openness at Universities*, 6 February (Washington, DC: NRC). Also set up was the Committee on the Science and Technology Agenda for Countering Terrorism.

4. Committee on Research Standards and Practices to Prevent the Destructive Application of Biotechnology. 2003. *Biotechnology Research in an Age of Terrorism* (Washington, DC: National Research Council): 14–15.
5. G. Fink. 2003. 'Preface', in *Biotechnology Research in an Age of Terrorism* (Washington, DC: National Research Council): vii–viii.
6. Ibid.: vii.
7. Committee on Research Standards and Practices to Prevent the Destructive Application of Biotechnology. 2003. *Biotechnology Research in an Age of Terrorism* (Washington, DC: National Research Council): 1.
8. Ibid.: 8.
9. Ibid.: 18.
10. Ibid.: 2.
11. R. Jackson, A. Ramsay, C. Christensen, S. Beaton, D. Hall, and I. Ramshaw. 2001. 'Expression of Mouse Interleukin-4 by a Recombinant Ectromelia Virus Suppresses Cytolytic Lymphocyte Responses and Overcomes Genetic Resistance to Mousepox', *Journal of Virology* 75, 3: 1205–10.
12. C. Cello, A. Paul and E. Wimmer. 2002. 'Chemical Synthesis of Poliovirus cDNA: Generation of Infectious Virus in the Absence of Natural Template' *Science* 297: 1016–18.
13. A. M. Rosengard, Y. L. Zhiping Nie, and R. Jimenez. 2002. 'Variola Virus Immune Evasion Design' *PNAS*, 11 April.
14. Committee on Research Standards and Practices to Prevent the Destructive Application of Biotechnology. 2003. *Biotechnology Research in an Age of Terrorism* (Washington, DC: National Research Council): 86.
15. Ibid.: 13.
16. Ibid.: 8.
17. Ibid.: 31.
18. See R. Atlas, K. Berns, G. Cassell, and J. Shoemaker. 1997. *Preventing the Misuse of Microorganisms: The Role of the ASM in Protecting Against Biological Weapons* (Washington, DC: ASM); and S. Wright. 1994. *Molecular Politics* (Chicago: Chicago University Press).
19. Department of Defense. 1997 'Technical Annex', in *Proliferation: Threat and Response* (Washington, DC: US Department of Defense). See: http:// www.defenselink.mil/pubs/prolif97/annex.html
20. See *Emerging and Infectious Diseases* 5, 4: http://www.cdc.gov/ncidod/eid/vol5no4/contents.htm; and the Second National Symposium on Medical and Public Health Response to Bioterrorism: http://www.upmc-biosecurity.org/pages/events/2nd_symposia/presentations.html. For similar discussions, see 'Bio-Technology and Bio-Weapons: Weapon of the 21st Century?' 2001. American Association for the Advancement of Science Annual Meeting and Science Innovation Exposition, 15–20 February (San Francisco, CA), at: http://www.aaas.org.

21. C. M. Fraser and M. Dando. 2001. 'Genomics and Future Biological Weapons: The Need for Preventive Action by the Biomedical Community', *Nature Genetics*, 22 October. This article was received at *Nature Genetics* on 5 September. Though, as a notable exception to general practice, see M. Cho, D. Magnus, A.L. Caplan, D. McGee, and the Ethics of Genomics Group. 1999. 'Ethical Considerations in Synthesizing a Minimal Genome', *Science* 286, 5447: 2087–90.
22. For a discussion of various biosecurity controls, see G. Epstein. 2001. 'Controlling Biological Warfare Threats', *Critical Reviews in Microbiology* 27, 4: 321–54.
23. See, for instance, The Lancet Infectious Diseases. 2001. 'Too Much Learning is a Dangerous Thing', *The Lancet Infectious Diseases* 1: 287; J. Gannon. 2001. 'Viewing Mass Destruction through a Microscope', *New York Times*, 11 October; and A. Pollack. 2001. 'Scientists Ponder Limits to Address Risk of Data Aiding Bioweapon Design', *New York Times*, 27 November.
24. C. Dennis. 2001. 'The Bugs of War', *Nature* 411 (22 October): 232–5.
25. Ibid.: 235.
26. Ibid.: 235.
27. P. Aldhous. 2001. 'Biologists Urged to Address Risk of Data Aiding Bioweapon Design', *Nature* 414 (15 November): 237–8.
28. A sentiment echoed elsewhere, such as in S. Morse. 2003. 'Bioterror R&D'. Presented at Meeting on National Security and Research in the Life Sciences National Academies and the Center for Strategic and International Studies, 9 January (Washington, DC).
29. G. Poste. 2001. 'A Rude Awakening to the Forces of Good and Evil', *Financial Times*, 27 November: insert iv.
30. For detail of a symposium about these matters, see American Association for the Advancement of Science. 2001. *Symposium Report on The War on Terrorism: What Does it Mean for Science?* (Washington, DC: AAAS). See: http:// www.aaas.org/spp/scifree/terrorism/report.shtml
31. G. Epstein. 2001. 'Controlling Biological Warfare Threats', *Critical Reviews in Microbiology* 27, 4: 321–54 at 336.
32. Ibid.: 337.
33. Central Intelligence Agency. 2003. *The Darker Bioweapons Future*, 3 November (Arlington, VA: Office of Transnational Issues); and B. Smith, T. Inglesby, Thomas and T. O'Toole. 2003. 'Biodefense R&D', *Biosecurity and Bioterrorism* 1, 3: 193–202.
34. See R. A. Zilinskas and J. B. Tucker. 2002. 'Limiting the Contribution of the Open Scientific Literature to the Biological Weapons Threat', *Journal of Homeland Security* (December). See: http://www.homelandsecurity.org/journal/Articles/tucker.html
35. Ibid.
36. R. Carlson. 2003. 'The Pace and Proliferation of Biological Technologies', *Biosecurity & Bioterrorism* 1, 3: 203–14.
37. For instance, in addition the other publications referenced elsewhere, see J. Stern. 2002/3. 'Dreaded Risks and the Control of biological Weapons', *International Security* 27, 3: 89–123; B. Vastag. 2003. 'Openness in Biomedical Research Collides With Heightened Security Concerns', *JAMA* 289: 686–90; K. Nixdorff and W. Bender. 2002. 'Ethics of University Research, Biotechnology

and Potential Military Spin-Off', *Minerva* 40: 15–35; The US National Academies and the Center for Strategic and International Studies Meeting on *National Security and Scientific Openness*, 9 January 2003 (Washington, DC); R. M. Atlas. 2002. 'National Security and the Biological Research Community', *Science* 298: 753; D. Shea. 2003. *Balancing Scientific Publication and National Security Concerns*, 10 January (Washington, DC: Congressional Research Service); The Sunshine Project. 2003. *Emerging Technologies: Genetic Engineering and Biological Weapons* (Austin, TX: The Sunshine Project); and Committee on Science, House of Representatives (US). 2002. *Background Paper to Conducting Research During the War on Terrorism*, 10 October. Available at: http://www.house.gov/science.

38. J. Couzin. 2002. 'A Call for Restraint on Biological Data', *Science* 297: 749–50.
39. S. Block 2002. 'A Not-So-Cheap Stunt', *Science* 297, 5582: 769–70.
40. D. Kennedy. 2002. 'Response to "A Not-So-Cheap Stunt" ', *Science* 297, 5582: 770.
41. A. Müllbacher and M. Logbis. 2001. 'Creation of Killer Poxvirus Could Have Been Predicted', *Journal of Virology* (September): 8353–5.
42. Homeland Security Act of 2002 Sec. 892(a)(1)(B).
43. See E. Check. 2002. 'US Prepares Ground for Security Clampdown', *Nature* 418: 906; and G. Knezo. 2003. *'Sensitive but Unclassified' and Other Federal Security Controls on Scientific and Technical Information*, 2 April (Washington, DC: Congressional Research Service).
44. M. Ensirenk. 2002. 'Entering the Twilight Zone of What Material to Censor', *Science*, 22 November.
45. Journal Editors and Authors Group. 2003. *PNAS*, 100, 4: 1464. See as well E. Chech. 2003. 'US Officials Urge Biologists to Vet Publication for Bioterror Risk', *Nature*, 16 January: 197; and J. Couzin. 2002. 'A Call for Restraint on Biological Data', *Science*, 2 August: 749–50.
46. See as well D. Kennedy. 2002. 'Two Cultures', *Science* 299 (21 February): 148; R. M. Atlas. 2002. 'National Security and the Biological Research Community', *Science* 298: 753; C. M. Vest. 2003. 'Balancing Security and Openness in Research and Education', *Academe*, September/October; and Committee on Science, House of Representatives (US). 2002. *Background Paper to Conducting Research During the War on Terrorism*, 10 October. Available at: http://www.house.gov/science.
47. A. Salyers. 2002. 'Science, Censorship, and Public Health', *Science* 296, 5568: 617.
48. Ibid.
49. R. Atlas. 2003. 'Preserving Scientific Integrity and Safeguarding Our Citizens'. Presented at Meeting on National Security and Research in the Life Sciences National Academies and the Center for Strategic and International Studies, 9 January (Washington, DC).
50. B. Alberts. 2002. 'Engaging in a Worldwide Transformation: Our Responsibility as Scientists for the Provision of Global Public Goods', Annual Meeting of the National Academy of Sciences, 29 April (Washington, DC. Emphasis in original.
51. J. Marburger. 2003. 'Perspectives on Balancing National Security and Openness in the Life Sciences'. Presented at Meeting on National Security and Research in the Life Sciences National Academies and the Center for Strategic and International Studies, 9 January (Washington, DC).

52. Committee on Genomics Databases for Bioterrorism Threat Agents. 2004. *Seeking Security* (Washington, DC: National Research Council): 7.
53. M. Leitenberg. 2001. 'Biological Weapons in the Twentieth Century', *Critical Reviews in Microbiology* 27, 4: 267–320.
54. J. Marburger. 2003. Address to the 2004 Annual Biodefense Conference, 27–28 August (Washington, DC).
55. L. Wells. 2003. 'Policies and Prospects'. Presented at Meeting on National Security and Research in the Life Sciences National Academies and the Center for Strategic and International Studies, 9 January (Washington, DC).
56. As part of contributing to debates about various biosecurity initiatives in the United States, both the American Association for the Advancement of Science: (http://www.aaas.org/spp/post911/); and the ASM: (http:// www.asm.org/Policy/index.asp?bid=520) established specific web sites which contain a range of analyses.
57. P. Albright. 2003. ' "Sensitive" Information in the Life Sciences'. Presented at Meeting on National Security and Research in the Life Sciences National Academies and the Center for Strategic and International Studies, 9 January (Washington, DC).
58. Central Intelligence Agency. 2003. *The Darker Bioweapons Future*, 3 November (Arlington, VA: Office of Transnational Issues).
59. D. Kennedy. 2002. 'Balancing Terror and Freedom', *Science*, 13 December: 2091.
60. For an overview of this initiated program of research, see White House. 2004. *Biodefense for the 21st Century*, 28 April (Washington, DC: US White House).
61. For an analysis of funding figures, see A. Schuler. 2005. 'Billions for Biodefense', *Biosecurity and Bioterrorism* 3, 2: 94–101.
62. E. Choffnes. 2002. 'Bioweapons: New Labs, More Terror?', *Bulletin of the Atomic Scientists* 58, 5: 28–32; and J. Knight. 2002. 'Biodefence Boost Leaves Experts Worried over Laboratory Safety', *Nature*, 14 February.
63. Health and Human Services. 2004. *HHS will lead Government-wide Effort to Enhance Biosecurity in 'Dual use' Research*, 4 March (Washington, DC: HHS Press Office).
64. H. O. Smith, C. A. Hutchison, C. Pfannkoch and J. C. Venter. 2003. 'Generating a Synthetic Genome by Whole Genome Assembly', *PNAS* 100, 26: 15440–5.
65. D. MacKenzie. 2003. 'U.S. Develops Lethal New Viruses' *New Scientist* 30 October.
66. L. M. Wein and Y. Liu. 2005. 'Analyzing a Bioterror Attack on the Food Supply', *PNAS* 102, 28: 9984–9.
67. L. M. Wein. 2005. 'Got Toxic Milk?', *New York Times*, 30 May.
68. See, e.g., S. Wright. 2004. 'Taking Biodefense too Far', *Bulletin of the Atomic Scientists* 60, 6: 58–66; and M. Leitenberg and G. Smith. 2005. *'Got Toxic Milk?': A Rejoinder*, 17 June (Washington, DC: Center for Arms Control and Non-Proliferation).
69. P. Campbell. 2005. 'Dual-Use Biomedical Research and the Roles of Journals', The InterAcademy Panel on International Issues, the International Council for Science, the InterAcademy Medical Panel and The National Academies of the United States, *International Forum on Biosecurity*, 20 March (Como, Italy).
70. B. Alberts and H. V. Fineberg. 2004. 'Harnessing New Science is Vital for Biodefense and Global Health', *PNAS* 101, 31: 11177.

71. J. Petro, T. Plasse, and J. McNulty. 2003. 'Biotechnology: Impact on Biological Warfare and Biodefense', *Bioterrorism and Biosecurity* 1, 3: 161–8.
72. J. Turner. 2004. *Beyond Anthrax: Confronting the Future Biological Weapons Threat*, May (Washington, DC). See as well M. Williams. 2003. 'The Looming Threat', *Acumen* 1, IV): 40–50.
73. See M. Enserink and J. Kaiser. 2005. 'Has Biodefense Gone Overboard?', *Science* 307, 5714: 1396–8; M. Leitenberg, J. Leonard, and R. Spertzel. 2004. 'Biodefense Crossing the Line', *Politics and the Life Sciences* 22, 2: 1–2; and D. Fidler. 2005. *Balancing Public Health and Bioterrorism Defenses*, 24 March (Washington, DC: Center for Arms Control and Non-Proliferation).
74. American Medical Association. 2005. *Guidelines to Prevent Malevolent Use of Biomedical Research*, CEJA Report 9-A-04 (Chicago: AMA Council on Ethical and Judicial Affairs); World Health Organization. 2005. *Governing Life Science Research - Opportunities and Risk for Public Health* (Paris: WHO); BBSRC, Medical Research Council and Wellcome Trust. 2005. *Managing the Risks of Misuse Associated with Grant Funding Activities* (London: BBSRC, MRC, Wellcome Trust). See as well E. Zerhouni. 2005. *Statement from Third Meeting of the National Science Advisory Board for Biosecurity*, 21 November (Bethesda, MD).
75. B. Rappert. 2004. 'National Security, Terrorism & Life Science Research', *Science and Technology Policies for the Anti-terrorism Era*, 12–14 September (Manchester, UK); and B. Rappert. 2004. 'Dual Use Research as a Social Problem', *XVI Amaldi Conference on Problems of Global Security*, 18–20 November.
76. As in S. Herrera. 2005. 'Preparing the World for Synthetic Biology', *Technology Review*, January; G. Chruch. 2004. 'Let us Go Forth and Safely Multiply', *Nature*, 24 November; and *The First International Meeting on Synthetic Biology*, 10–12 June 2004 (Cambridge, MA). See: http://web.mit.edu/ synbio/release/ conference/
77. J. Steinbruner, E. D. Harris, N. Gallagher, and S. Gunther. 2003. *Controlling Dangerous Pathogens: A Prototype Protective Oversight System*, 5 February, available at: www.puaf.umd.edu/cissm/projects/amcs/pathogens.html.
78. Though see M. Lim. 2005. 'Hostile Use of the Life Science', *New England Journal of Medicine* 353: 2214–15.
79. A. Jamrozik and L. Nocella. 1998. *The Sociology of Social Problems* (Cambridge: Cambridge University Press): 8.
80. See M. Spector and J. Kitsuse. 1977. *Constructing Social Problems* (Menlo Park, CA: Cummings); and J . Holstein and G. Miller (eds). 1993. *Reconsidering Social Constructionism* (New York: Aldine de Gruyter).
81. For a treatment of relativism in realism, see D. Edwards, M. Ashmore, and J. Potter. 1995. 'Death and Furniture', *History of the Human Sciences* 8: 26–35.
82. J. Holstein and G. Miller. 1993. 'Social Constructionism and Social Problems Work', in J. Holstein and G. Miller (eds), *Reconsidering Social Constructionism* (New York: Aldine de Gruyter): 151; and J. Wilmoth and P. Ball. 1995. 'Arguments and Action in the Life of a Social Problem', *Social Problems* 42, 3: 318–43.
83. S. Woolgar and D. Pawluch. 1985. 'Onotological Gerrymandering', *Social Problems* 32, 3: 214–27.
84. P. Conrad. 1997. 'Public Eyes and Private Genes', *Social Problems* 44, 2: 139–54: 159.

85. P. Ibarra and J. Kituse. 1993. 'Vernacular Constituents of Moral Discourse', in J. Holstein and G. Miller (eds), *Reconsidering Social Constructionism* (New York: Aldine de Gruyter).
86. J. Schneider. 1993. ' "Members Only" ', in J. Holstein and G. Miller. (eds) *Reconsidering Social Constructionism* (New York: Aldine de Gruyter).
87. For a discussion of such issues, see *Social Problems* 39, 1: 35–9; and A. Gordon. 1993. 'Twenty-Two Theses on Social Constructionism', in J. Holstein and G. Miller (eds), *Reconsidering Social Constructionism* (New York: Aldine de Gruyter).
88. R. Àlvarez. 2001. 'The Social Problem as an Enterprise', *Social Problems* 48, 1: 3–4.
89. J. Best. 1993. 'But Seriously Folks', in J. Holstein and G. Miller (eds), *Reconsidering Social Constructionism* (New York: Aldine de Gruyter).
90. R. M. Atlas. 2002. 'National Security and the Biological Research Community', *Science* 298: 753.

Chapter 2 Discussing Science

1. J. Marburger. 2003. 'Perspectives on Balancing National Security and Openness in the Life Sciences'. Presented at Meeting on National Security and Research in the Life Sciences National Academies and the Center for Strategic and International Studies, 9 January (Washington, DC).
2. S. Boehlert. 2002. Background Paper to Committee on Science, House of Representatives Hearing Conducting Research During the War on Terrorism, 10 October (US). Available at: http://www.house.gov/science
3. G. Epstein. 2001. 'Controlling Biological Warfare Threats', *Critical Reviews in Microbiology* 27, 4: 337.
4. D. Shea. 2003. *Balancing Scientific Publication and National Security Concerns*, 10 January (Washington, DC: Congressional Research Service): 3.
5. Committee on Genomics Databases for Bioterrorism Threat Agents. 2004. *Seeking Security* (Washington, DC: National Research Council): 7. For other references to norms, see G. Knezo. 2002. *Counter Terrorism* (New York: Novinka); and Policy and Global Affairs Division. 2002. *Post-September 11 Scientific Openness at Universities*, 6 February (Washington, DC: The National Academies). Also it should be noted that norms have not just been taken as given and under threat in biosecurity discussions. Rather, the building of 'cultural norms' of responsibility has been taken as a key aim of many policy initiatives. As in E. Zerhouni. 2005. *Statement from Third Meeting of the National Science Advisory Board for Biosecurity*, 21 November (Bethesda, MD).
6. R. Merton. [1942]1973. 'The Normative Structure of Science', in N. Storer (ed.), *The Sociology of Science* (Chicago: University of Chicago Press).
7. For a vivid illustration of the importance of this way of thinking, see the 21 November 2005 NSABB meeting, available at: http://www.biosecurityboard.gov/meetings.asp.
8. See, for instance, R. Eisenberg. 1996. 'Intellectual Property Issues in Genomics', *Trends Biotechnology* 14, 8: 302–7; The Lancet. 2005. 'Financial Ethics Pit NIH Scientists against Government', *The Lancet*, 13–19 August: 537–8; K. Packer. 1994. 'Academic-Industry Relations Selected Bibliography', *Science and Public Policy* 21: 117–19; and D. Blumenthal. 1996. 'Ethics Issues

in Academic-Industry Relationships in the Life Sciences', *Academic Medicine* 12: 1291–6.

9. National Academy of Science. 1997. *Intellectual Property Rights and Research Tools in Molecular Biology* (Washington, DC: National Academy Press); Nature. 2005. 'Experts say US Authorities should Change Patent Laws', *Nature* 438: 409; and Royal Society. 2003. *Keeping Science Open* (London: Royal Society).
10. J. Ziman. 2000. *Real Science* (Cambridge: Cambridge University Press); and H. Etzkowitz. 1989. 'Entrepreneurial Science in Academy', *Social Problems* 36, 1: 14–29.
11. S. Krimsky. 2003. *Science in the Private Interest* (Lanham, MD: Rowman & Littlefield): 1.
12. Ibid.: 3.
13. M. MacKenzie, P. Keating, and A. Cambrosio. 1990. 'Patents and Free Scientific Information in Biotechnology', *Science, Technology, and Human Values* 15, 1: 65–83.
14. Following a more constructivist line of analysis, later they revised their interest in this topic to considering how scientific findings and materials are able to gain the status of public goods. See A. Cambrosio and P. Keating. 1998. 'Monoclonal Antibodies', in A. Thackray (ed.), *Private Science* (Philadelphia: University of Pennsylvania Press).
15. For discussion of rationales, see P. Shorett, P. Rabinow, and P. Billings. 2003. 'The Changing Norms of the Life Sciences', *Nature Biotechnology* 21: 123–5; and S. Hill and T. Turpin 1993. 'The Clashing of Academic Symbols', *Science as Culture* (February).
16. R. K. Merton. 1969. 'Behavior Pattern of Scientists', *American Scientist* 57: 1–23.
17. I. Mitroff. 1974. *The Subjective Side of Science* (Amsterdam: Elsevier).
18. M. Mulkay. 1975. 'Norms and Ideology in Science', *Social Science Information* 15, 4–5: 637–56.
19. For a somewhat parallel argument about the relation of science to war in the United Kingdom, see D. Edgerton. 1996. 'British Scientific Intellectuals and the Relations of Science and War in Twentieth Century Britain', in P. Forman and J. M. Sanchez Ron (eds), *National Military Establishments and the Advancement of Science: Studies in Twentieth Century History* (Dordrecht: Kluwer).
20. Say, with regard to the public confidence this inspires in members of the public.
21. N. Gilbert and M. Mulkay. 1984. *Opening Pandora's Box* (Cambridge: Cambridge University Press).
22. Ibid.: 56.
23. M. Mulkay. 1985. *The Word and the World* (London: Allen & Unwin): ch. 4. Much of this chapter centered on earlier claims about replication made in H. Collins. 1975. 'The Seven Sexes', *Sociology* 9, 2: 205–24.
24. S. Cunningham-Burely and A. Kerr. 1999. 'Defining the "Social" ', *Sociology of Health & Illness* 21, 5: 647–68; and A. Kerr, S. Cunningham-Burely, S. and A. Amos. 1997. 'The New Genetics: Professionals' Discursive Boundaries', *The Sociological Review*: 280–302.
25. For a background discussion of boundaries, see T. Gieryn. 1995. 'Boundaries of Science', in S. Jasanoff, G. Markle, J. Petersen, and T. Pinch (eds), *Handbook of Science and Technology Studies* (London: Sage Publications); and T. Gieryn. 1999. *Cultural Boundaries of Science* (Chicago: University of Chicago Press).

26. H. Collins. 2004. *Gravity's Shadow* (London: Chicago University Press).
27. H. Collins. 1998. 'The Meaning of Data', *American Journal of Sociology* 104, 2: 293–337.
28. It follows from this that achieving a relevant analysis for Collins requires a reasonable technical understanding of the field in question; one that does not qualify the social analyst as a scientist but allows this person to understand the basis for claims.
29. H. Collins. 2004. *Gravity's Shadow* (London: Chicago University Press): 756.
30. For further development of this theme, see K. Knorr Cetina. 1999. *Epistemic Cultures: How the Sciences Make Knowledge* (Cambridge, MA: Harvard University Press); and A. Pickering. 2006. 'Culture', in T. Bennett and J. Frow (eds), *Handbook of Cultural Analysis* (London: Sage).
31. R. Kurzweil and B. Joy. 2005. 'More Spanish Flu', *New York Times*, 17 October.
32. J. Callen. 2005. Letter to the Editor, *New York Times*, 20 October.
33. J. Aach. 2005. Letter to the Editor, *New York Times*, 20 October.
34. K. Tzamarot. 2005. Letter to the Editor, *New York Times*, 20 October.
35. A. Casadevall. 2005. 'Issues of Relevance to Criteria Development'. Presentation to NSABB Meeting, 30 June (Bethesda, MD).
36. T. Shenk. 2003. 'Sensitive Research'. Presented at Meeting on National Security and Research in the Life Sciences National Academies and the Center for Strategic and International Studies, 9 January (Washington, DC).
37. G. Epstein. 2001. 'Controlling Biological Warfare Threats', *Critical Reviews in Microbiology* 27, 4: 321–54, at 336.
38. As in A. Müllbacher and M. Logbis. 2001. 'Creation of Killer Poxvirus Could have been Predicted', *Journal of Virology*, September: 8353–5.
39. D. MacKenzie. 2005. 'US Develops New Lethal Viruses', *New Scientist*, 1 November.
40. Statement made at the conference titled 'Do No Harm', London, The Royal Society, 7 October 2004.
41. S. Jasanoff. 1997. 'Civilization and Madness', *Public Understanding of Science* 6: 221–32, available at: http://pus.sagepub.com/content/vol6/issue3/
42. B. Rappert. 2003. 'Coding Ethical Behaviour: The Challenges of Biological Weapons', *Science & Engineering Ethics* 9, 4: 453–70.
43. For a discussion of this, see H. Collins. 1998. 'The Meaning of Data', *American Journal of Sociology* 104, 2: 293–337.

Chapter 3 Inquiry, Engagement, and Education

1. For instance, see Report of Royal Society and Wellcome Trust Meeting. 2004. 'Do No Harm – Reducing the Potential for the Misuse of Life Science Research', 7 October; Royal Society. 2002. *Submission to the Foreign and Commonwealth Office Green Paper on Strengthening the Biological and Toxin Weapons Convention*, September (London: Royal Society); and World Medical Association. 2002. *Declaration of Washington on Biological Weapons* (Washington, DC: WMA).
2. See, e.g., P. Aldhous. 2001. 'Biologists Urged to Address Risk of Data Aiding Bioweapon Design', *Nature* 414 (15 November): 237–8.
3. Royal Society. 2002. *Submission to the Foreign and Commonwealth Office Green Paper on Strengthening the Biological and Toxin Weapons Convention*, September: 4.

4. Report of Royal Society and Wellcome Trust Meeting 2004. 'Do No Harm – Reducing the Potential for the Misuse of Life Science Research', 7 October: 1.
5. World Medical Association. 2002. *Declaration of Washington on Biological Weapons* (Washington, DC: WMA).
6. National Research Council. 2003. *Biotechnology Research in and Age of Terrorism* (Washington, DC: National Academies Press).
7. As evident from the NAS and ASSS meeting 'Education and Raising Awareness: Challenges for Responsible Stewardship of Dual Use Research in the Life Sciences', 8–9 September 2005 (Washington, DC).
8. M. Billig, S. Condo, D. Edwards, M. Gane, D. Middleton, and A. Radley. 1989. *Ideological Dilemmas* (London: Sage).
9. And to the extent scientists and security analysts are supposed to work in partnership to define the nature of threats, then the occasions in which questions are posed of 'who really knows?' are likely to multiply.
10. For what little information there is on this matter, see W. Barnaby. 1997. *The Plague Makers* (London: Vision).
11. C. McLeish and P. Nightingale. 2005. *Effective Action to Strengthen the BTWC Regime: The Impact of Dual Use Controls on UK Science*, Bradford Briefing Paper No. 17, May.
12. B. Rappert and M. Dando. Economic and Social Research Council Award 'Accountability and the Governance of Expertise: Anticipating Genetic Bioweapons' Project Ref: L144250029.
13. To explain this choice, in relation to muscarinic acetylcholine receptors, nerve agents developed in the early twentieth century such as tabun, sarin and VX functioned by inhibiting acetylcholinesterase. Acetylcholine is normally broken down in the synaptic cleft by an enzyme called acetylcholinesterase. Past nerve agents acted by inhibiting the function of acetylcholinesterase. Since acetylcholine has a significant role in both the central and peripheral nervous systems, the net result is total disruption of their functioning. In the search for treatments to neurodegenerative diseases such as Alzheimer's and Parkinson's diseases, major attempts have been made in recent decades to specify the functioning of acetylcholine and its receptors, such as muscarinic receptors. The latter have been found to be involved in motor control, temperature regulation, cardiovascular regulation, and memory. Recently the use of 'knock-out' mice and other techniques has enabled a greater understanding of the behavioral effects of eliminating the genes for individual muscarinic receptor sub-types. In relation to bioweapons, such developments may enable both the more effective targeting of acetylcholine and the ability to achieve specific effects (e.g., incapacitation).
14. J. Dewey. 1929. *The Quest for Certainty* (London: George Allen & Unwin): 233.
15. M. Cochran. 2002. 'Deweyan Pragmatism and Post-Positivist Social Science in IR', *Millennium* 31, 3: 527.
16. Ibid.: 530.
17. T. Nickles. 1988. 'Questioning and Problems in Philosophy of Science', in M. Meyer (ed.), *Questions and Questioning* (Berlin: Walter de Gruyter).
18. R. Rorty. 1999. *Philosophy and Social Hope* (London: Penguin): xxv.
19. R. Rorty, 1979. *Philosophy and the Mirror of Nature* (Princeton, NJ: Princeton University Press): 377.

20. Ibid: 378.
21. R. Rorty. 1999. *Philosophy and Social Hope* (London: Penguin): 34.
22. Ibid.: ch. 3.
23. E.g., in relation to the 'traits of experimental inquiry', see M. Cochran. 1999. *Normative Theory in International Relations* (Cambridge: Cambridge University Press): ch. 6.
24. C. Argyris. 1994. *On Organizational Learning,* 2nd edn (London: Blackwell): 131.
25. C. Argyris. 2006. *Reasons and Rationalizations* (Oxford: Oxford University Press): 212.
26. C. Argyris, R. Putman. and D. M. Smith. 1985. *Action Science* (London: Jossey-Bass): 230.
27. C. Argyris and D. Schön 1996. *Organizational Learning II* (London: Addison Wesley): 79.
28. Indeed, it would be difficult to overplay how ever present and engrained Argyris regards defensive, Model I types of reasoning. This leads to a rather pessimistic conclusion about the likelihood for creating futures in which systematic and costly error is not present.
29. C. Argyris. 2006. *Reasons and Rationalizations* (Oxford: Oxford University Press): 134–42.
30. C. Argyris and D. Schön 1996. *Organizational Learning II* (London: Addison Wesley): 80–4.
31. Ibid: 82–3.
32. C. Argyris. 2003. 'A Life Full of Learning', *Organizational Studies* 24, 7: 1178–92.

Chapter 4 Learning to Respond

1. A. Johnson. 1996. 'It's Good to Talk', *Sociological Review* 44: 517–36; and L. Chiu. 2003. 'Transformational Potential of Focus Group Practice in Participatory Action Research', *Action Research* 1: 165–83.
2. D. Stewart and P. Shamdasani. 1992. *Focus Groups* (London: Sage).
3. J. Kitzinger and R. Barbour. 1999. 'Introduction', in R. Barbour and Jenny Kitzinger (eds), *Developing Focus Group Research* (London: Sage).
4. D. Morgan. 1998. *Focus Groups as Qualitative Research* (London: Sage): 12; de-italicized quote.
5. R. Krueger. 1998. *Developing Questions for Focus Groups* (London: Sage).
6. C. Farquhar and R. Das. 1999. 'Are Focus Groups Suitable for "Sensitive" Topics?', in R. Barbour and J. Kitzinger (eds), *Developing Focus Group Research* (London: Sage); and J. Kitzinger. 1994. 'The Methodology of Focus Groups', *Sociology of Health & Illness* 16, 1: 103–21.
7. R. Baker and R. Hinton. 1999. 'Do Focus Groups Facilitate Meaningful Participation in Social Research?', in R. Barbour and J. Kitzinger (eds), *Developing Focus Group Research* (London: Sage).
8. B. Thayer-Bacon. 2003. 'Pragmatism and Feminism as Qualified Relativism', *Studies in Philosophy and Education* 22: 419.
9. K. O'Brien. 1993. 'Improving Survey Questionnaires', in D. Morgan (ed.), *Successful Focus Groups* (London: Sage).
10. T. Albrecht., G. Johnson, and J. Walther. 1993. 'Understanding Communication Processes in Focus Groups', in D. Morgan (ed.), *Successful Focus Groups* (London: Sage).

11. Defining a research protocol, conducting disciplined moderation, and establishing feedback between researchers and group participants, other researchers and outside experts are all presented as vital for producing systematic results. See R. Krueger. 1998. *Developing Questions for Focus Groups* (London: Sage).
12. J. Kitzinger. 1994. 'The Methodology of Focus Groups', *Sociology of Health & Illness* 16, 1: 103–21.
13. D. Morgan. 1993. *Successful Focus Groups* (London: Sage).
14. For instance, M. Scott and S. Lyman. 1968. 'Accounts', *American Sociological Review* 33: 46–62.
15. D. Morgan. 1998. *Focus Groups as Qualitative Research* (London: Sage): 25.
16. D. Edwards. 1997. *Discourse and Cognition* (London: Sage); and D. Silverman (ed.). 2004. *Qualitative Research* (London: Sage).
17. C. Waterton and B. Wynne. 1999. 'Can Focus Groups Access Community Views?', in R. Barbour and J. Kitzinger (eds), *Developing Focus Group Research* (London: Sage): 127.
18. D. Morgan. 1998. *Focus Groups as Qualitative Research* (London: Sage): 58.
19. J. Kitzinger. 1994. 'The Methodology of Focus Groups', *Sociology of Health & Illness* 16, 1: 106.
20. J. Kitzinger and R. Barbour. 1999. 'Introduction', in R. Barbour and J. Kitzinger (eds), *Developing Focus Group Research* (London: Sage): 5.
21. R. Krueger. 1998. *Developing Questions for Focus Groups* (London: Sage).
22. D. Morgan. 1998. *Focus Groups as Qualitative Research* (London: Sage): 81.
23. Originally the seminars were going to be held as part branch meetings of the Institute of Biology, a professional body representing biologists in the United Kingdom. Despite a letter of endorsement sent out from the Executive Director and publicity within the Institute's newsletter, in the end, attempts to get members interested in attending and organizing such meetings proved resoundingly unsuccessful.
24. After each session we, as facilitators, debriefed to identify themes and concerns. Based on further examination of written and audio data, after a group of seminars (numbers noted below) we wrote up an analysis of that group. This was sent back to local organizers for distribution to participants and feedback.
25. One audio proved largely inaudible and so the analysis of it in this book is based on what can be heard plus notes taken during and after the session.
26. Rather than speculating whether the responses offered were given out of concerns about group acceptability, personal antagonisms, or other motivating factors, we encouraged a general questioning of the justifications for statements. In other words, the concern is not so much with whether responses are by some metric authentic or biased, but rather with treating accounts in their own right and finding ways of testing the basis for whatever is said in the service of promoting mutual understanding and further reflection.
27. H. Smith, C. Hutchison, C. Pfannkoch, and C. Venter. 2003. 'Generating a Synthetic Genome by Whole Genome Assembly', *PNAS* 100: 15440–5.
28. See, e.g., A. Regalado. 2005. 'Biologist Venter Aims to Create Life from Scratch', *The Wall Street Journal*, 29 June.
29. In the last grouping of seminars we used our acquired knowledge of the dilemmatic aspects of communication as a base for assembling a range of devil's advocate 'counter-points', which were meant to enable further

consideration of the assumptions and inferences supporting answers to the questions raised.

30. Such as: How can we encourage due consideration of the possible consequences of the misuse of research? How can we promote the proper use of science-based activities and knowledge and encourage appropriate oversight of such work? Is it necessary to provide guidance on how to deal with research that throws up unexpected or unpredictable results of relevance to the BTWC prohibitions? How might we promote consideration among research and project funders of BTWC issues when considering proposals; e.g., whether the research could be misused in the future and what steps might help to prevent this?
31. World Medical Association. 2002. *Declaration of Washington on Biological Weapons* (Washington, DC: WMA).
32. J. Rudolph, S. Taylor, and E. Foldy. 2000. 'Collaborative Off-line Reflection, in P. Reason and H. Bradbury (eds), *Handbook of Action Research* (London: Sage): 405–12.

Chapter 5 Openness and Constraint

1. See K. Tracy. 1997. *Colloquium* (London: Ablex Publishing); and M. Billig, S. Condo, D. Edwards, M. Gane, D. Middleton, and A. Radley. 1989. *Ideological Dilemmas* (London: Sage).
2. See J. Kitzinger. 1994. 'The Methodology of Focus Groups', *Sociology of Health & Illness* 16,1: 103–21.
3. It should be noted, though, that any work entailing the weaponization of biological agents was steadfastly rejected.
4. Assuming the results eventually obtained were regarded as reasonably foreseeable from the start of the work.
5. August 2004.
6. As in their overwhelming use of an empiricist repertoire elaborated in Chapter 2, see N. Gilbert and M. Mulkay. 1984. *Opening Pandora's Box*. Cambridge: Cambridge University Press.
7. So, for instance, as in the exchange:

   P10: Also if you start to be selective about domains that you publish in you start to tighten assumptions about specific people and specific uses in those domains, and unless you have a policy of open publishing, you, those assumptions are never going to be right. I mean, why just because it's published in the scientific world that it's used for good rather than bad, who's to define what's good or bad? I mean.

   MD: Yes, so if I understand you what you're saying that the person with malevolent intent with a scientific background is not going to read *New Scientist*, they're going to read *Journal of Virology* and they're going to know what the implications of that kind of manipulation would be.

   P10: I just think it's dangerous to give anyone decision-making power about the domains which any information should be restricted to.

   (Seminar 6)

   As well as in:

   P9: But how can you be certain that some country that you don't want to know that, hasn't already discovered it and kept it secret to themselves and the rest of the world is completely ignorant, so how, who do you decide is going to know about it and not know about it, how much do you restrict it, do you just restrict it to Britain or should you restrict it to other? See what I mean, like, how do you know that a country that you're really scared will find out this information doesn't already know it, like some of these countries are rich, they hope they can carry out the same sort of research that we're carrying out and if they discover it they're going to keep it to themselves if they think no-one else knows about it so they can use it, if they want to use it in a bad way, yeah, anyway.

(Seminar 9)

8. In a similar vein of speaking to the mix of policy demands, scientists' actions and the inevitability of science, one participant argued:

   So I think a lot of policy work is quite sophisticated and I think the counterbalance to that is that scientists, in my experience, can be particularly adept at manipulating what they do to whatever policy area is identified, so under the heading [in your slide]: Scientific development, is it inevitable?, you could say well it shouldn't be inevitable because they've got to hit these criteria but scientists are increasingly very adept at showing how whatever research they are doing do, does hit, so I think by and large, in most areas not all, scientific development is inevitable and if you don't do it somebody else will, so we do use that as a convenient way of side-stepping our responsibility and saying y'know this is publicly funded research, if we're allowed to do it, if we hit the funding criteria we're going to do it and we will. If we could find a way round doing GM research in this country we would do it, wouldn't we? If we could do it within the regulations we'd certainly do it.

   (Seminar 7)

9. E. Schegloff, I. Koshik, S. Jacoby, and D. Olsher. 2002. 'Conversation Analysis and Applied Linguistics', *Annual Review Applied Linguistics* 22: 5.
10. See P. Drew and J. Heritage (eds). 1992. *Talk at Work* (Cambridge: Cambridge University Press).
11. J. Sinclair and R. Coulthard. 1975. *Towards an Analysis of Discourse* (Oxford: Oxford University Press).
12. N. Markee. 1995. 'Teachers' Answers to Students' Questions', *Issues in Applied Linguistics* 6, 2: 67.
13. A. McHoul. 1978. 'The Organization of Turns at Formal Talk in the Classroom', *Language and Society* 7: 188.
14. S. Clayman and J. Heritage. 2002. *The News Interview* (Cambridge: Cambridge University Press): 97.
15. See P. ten Have. 1999. *Doing Conversation Analysis* (London: Sage): 166–7.
16. S. Clayman and J. Heritage. 2002. *The News Interview* (Cambridge: Cambridge University Press): 13.
17. The closest thing to this was an interjection in one seminar noting the time limitations which mean the seminar had to wind to a close.
18. This in terms of suggesting the need to revisit a slide. Obviously references were made to information given in prior slides and their implication for the discussion at hand.

19. For a discussion of news interviews as co-constructed question and answer sessions, see S. Clayman and J. Heritage. 2002. *The News Interview* (Cambridge: Cambridge University Press).
20. See I. Hutchby and R. Wooffitt. 1998. *Conversation Analysis* I(Cambridge: Polity): ch. 3.
21. J. Heritage. 1985. 'Analysing News Interviews', in *Handbook of Discourse*, vol. 3 (London: Academic Press): 153.
22. I. Hutchby and R. Woofitt. 1998. *Conversation Analysis* (Cambridge: Polity): 167.
23. See as well K. Tracy. 1997. *Colloquium* (London: Ablex Publishing).
24. So this in contrast to the original slide question given in Box 4.1 of 'Should such experimental results have been widely circulated?'. This change was initiated after we noticed the differences in what widely circulated means.
25. S. Clayman and J. Heritage. 2002. *The News Interview* (Cambridge: Cambridge University Press): 113.

Chapter 6 Neutrality and Bias

1. D. Edwards. 1997. *Discourse and Cognition* (London: Sage).
2. J. Potter. 1997. *Representing Reality* (London: Sage); and J. Potter and M. Wetherell *Discourse and Social Psychology* (London: Sage).
3. E. Schegloff, I. Koshik, S. Jacoby, and D. Olsher. 2002. 'Conversation Analysis and Applied Linguistics', *Annual Review Applied Linguistics* 22: 5.
4. I. Mitroff. *The Subjective Side of Science* (Amsterdam: Elsevier).
5. The person I referred to as 'in the back there' was the one that made the comment that biodefense funding was a misallocation of resources because 'people are dying of AIDS and cancer'.
6. E. Schegloff, I. Koshik, S. Jacoby, and D. Olsher. 'Conversation Analysis and Applied Linguistics', *Annual Review Applied Linguistics* 22: 5. See as well S. Clayman and J. Heritage. 2002. *The News Interview* (Cambridge: Cambridge University Press): 235.
7. Of course, one benefit of focus group type interactions over more structured forms of research is that they afford significant opportunity for individuals to state what they find relevant.
8. In addition, as discussed in Chapter 1, it is not as if the author had no working assessments and presumptions about the issues discussed.
9. The lack of scrutiny was, though, in contrast to the post-seminar discussions which tended to be fairly free ranging.
10. S. Clayman and J. Heritage. 2002. *The News Interview* (Cambridge: Cambridge University Press): 151.
11. A. von Bubnoff. 2005. 'Special Report: The 1918 Flu Virus is Resurrected', *Nature* 437 (6 October): 794–5.

Chapter 7 Expertise and Equality

1. As in *An Open Letter* from Social Movements and other Civil Society Organizations to the Synthetic Biology 2.0 Conference, 20–22 May 2006

Berkeley, California. See: http://www.genewatch.org/article.shtml?als[cid]=507663&als[itemid]=537746
2. B. Rollin. 2006. *Science and Ethics* (Cambridge: Cambridge University Press).
3. Royal Society. 2006. *Science and the Public Interest* (London: Royal Society): 7.
4. New Scientist. 2001. 'Disaster in the Making', *New Scientist* 169, 2273: 4.
5. While Excerpts 7.6 and 7.7 both contained concerns about public perceptions of science given the coverage of it, one response combined both the need for greater public awareness and communicating with the need for frightening:

   P3: Well the kind of thing we've talked about is fairly obvious to most people we can all sit down most of us and write scenarios to make dreadful weapons, it's not difficult, you know it's the price you pay for your increased knowledge and understanding of the world, use it this way or that way. But if you go down the route of saying stop it, then you know, you'll be in even more serious trouble. I think the real problem is nothing to do with this, I think the problem is that the vast majority of the world's population has such a poor understanding of science that they can't evaluate things like that and therefore they're very readily frightened but I would argue that you actually want the population to be frightened, they should be frightened ((group laughter)) and they should be trying to persuade the politicians that we can deliver them when these weapons are required. I don't see any other solution really other than saying just stop science.

   (Seminar 16)
6. K. Tracy. *Colloquium* (London: Ablex Publishing): 90.
7. To be clear, certain statements were qualified by participants noting their specialized area of research was other than that seen as most relevant for the question posed. Such statements though were not statements that took issue with scientific expertise.
8. House of Commons Science and Technology Committee. 2003. *The Scientific Response to Terrorism* (London: HMSO).
9. As in and J. Kitzinger and R. Barbour. 1999. 'Introduction', in R. Barbour and J. Kitzinger (eds), *Developing Focus Group Research* (London: Sage); and D. Morgan. 1998. *Focus Groups as Qualitative Research* (London: Sage): 12.

Chapter 8 Closing Remarks

1. This is not to say though that there are not important concerns about partiality in questioning more widely. As mentioned in Chapter 6, Clayman and Heritage posed the question of how do news interviewers 'manage to assert themselves in an adversarial manner while maintaining a formally impartial or neutralist posture?' (see S. Clayman and J. Heritage. 2002. *The News Interview* (Cambridge: Cambridge University Press): 151).

Index

Note: Headings containing the wording 'research project' refer to the study carried out by the author and Malcolm Dando. Explanatory notes are indexed in the format page number:note number. References and non-substantive notes are not indexed.